GALILEO'S LOGICAL TREATISES

BOSTON STUDIES IN THE PHILOSOPHY OF SCIENCE

VOLUME 138

WILLIAM A. WALLACE

University of Maryland at College Park

GALILEO'S LOGICAL TREATISES

A Translation, with Notes and Commentary,
of His Appropriated Latin Questions on Aristotle's
Posterior Analytics

KLUWER ACADEMIC PUBLISHERS

DORDRECHT / BOSTON / LONDON

Library of Congress Cataloging-in-Publication Data

```
Galilei, Galileo, 1564-1642.
    [Selections. English. 1992]
    Galileo's logical treatises : a translation, with notes and
commentary, of his appropriated Latin questions on Aristotle's
Posterior analytics / William A. Wallace.
        p.   cm. -- (Boston studies in the philosophy of science ; v.
138)
    Includes bibliographical references and indexes.
    ISBN 0-7923-1578-2 (alk. paper). -- ISBN 0-7923-1579-0 (set : alk.
paper)
    1. Logic, Modern--16th century.   I. Wallace, William A.
II. Title.   III. Series.
Q174.B67  vol. 138
[B785.G22E5]
001'.01 s--dc20
[160]                                                            91-36925
ISBN 0-7923-1578-2
Set:  0-7923-1579-0
```

Published by Kluwer Academic Publishers,
P.O. Box 17, 3300 AA Dordrecht, The Netherlands.

Kluwer Academic Publishers incorporates
the publishing programmes of
D. Reidel, Martinus Nijhoff, Dr W. Junk and MTP Press.

Sold and distributed in the U.S.A. and Canada
by Kluwer Academic Publishers,
101 Philip Drive, Norwell, MA 02061, U.S.A.

In all other countries, sold and distributed
by Kluwer Academic Publishers Group,
P.O. Box 322, 3300 AH Dordrecht, The Netherlands.

Printed on acid-free paper

TABLE OF CONTENTS

TREATISE ON FOREKNOWLEDGES AND FOREKNOWNS

NOTES AND COMMENTARY

TREATISE ON DEMONSTRATION

NOTES AND COMMENTARY

LIST OF TABLES

PREFACE

As explained in the Latin Edition of Galileo's appropriated questions on Aristotle's *Posterior Analytics*, both that edition and its translation, here presented as *Galileo's Logical Treatises*, have been many years in preparation. For a variety of reasons the manuscript in which the treatises are preserved (MS Gal. 27) was never transcribed by the editor of Galileo's works, Antonio Favaro, and so was not included in the famous National Edition of 1890–1909. As a result the manuscript, itself written in Galileo's hand but in a cramped Latin that is difficult to read, has effectively lain dormant in Italian collections for some four hundred years. In 1964, however, in conjunction with his researches on the logic of Jacopo Zabarella (1533–1589), William F. Edwards examined the original at the Biblioteca Nazionale Centrale in Florence. His interest in Galileo's logic was kindled by the last question in the manuscript, that dealing with the demonstrative *regressus*, for its possible relationship to Zabarella and the scientific methodology developed at the University of Padua. Impressed by Galileo's account, Edwards decided to transcribe the manuscript in its entirety. Not until 1966, however, was he able to secure from the Biblioteca a microfilm of it that was suitable for his purposes. Working then from photostats, he completed most of the transcription by 1967, although a few residual problems remained. He continued to work on these until his move from the University of Buffalo to Emory University in Atlanta in 1970. Partly because of the move, partly because of differences between Galileo's and Zabarella's treatments of the *regressus* he was unable to explain, Edwards set his transcription aside and did not publish it at that time.

During roughly the same period, I also had begun work on another of Galileo's Latin manuscripts, the "Physical Questions" of MS Gal 46, which, unlike MS Gal. 27, Favaro had transcribed and included in the National Edition under the title of *Juvenilia* or youthful studies. In 1967 I made a sight translation of these questions into English, and for the next decade was occupied exploring the connections between them and the teachings of various Thomists in the late sixteenth century, including the Jesuits of the Collegio Romano. This led to my publication, ten years later, of *Galileo's Early Notebooks: The Physical Questions* (Notre Dame 1977), which contained a reworked translation of the manuscript along with an historical and paleographical commentary. I then put together a volume of studies documenting my supporting research on MS Gal. 46; this appeared some years later as *Prelude to Galileo: Essays on Medieval and Sixteenth-Century Sources of Galileo's Thought* (Reidel 1981).

Meanwhile, in 1972, while verifying Favaro's transcription of MS Gal. 46 against the autograph conserved in the Biblioteca Nazionale Centrale in Florence, I took the opportunity to examine the autograph of MS Gal. 27 and actually began to transcribe portions of it myself. My interest then was mainly one of checking the authorities cited by Galileo in that manuscript so as to compare them with those cited in MS Gal. 46. At that time I was corresponding with Alistair Crombie about Galileo's early Latin manuscripts, and he informed me that a transcription of MS Gal. 27 had already been made by one of his former students, Adriano Carugo. Crombie, in fact, arranged for Carugo to provide me with a copy of his transcription, and Carugo did so late in 1972, with the understanding that the work was for my private use only. From it I obtained the citation information I was seeking and appended it to a note of an essay I was then preparing on "Galileo and the Thomists." Upon my return to the U.S. at the end of 1972, I compared Carugo's transcription of MS Gal. 27 with a microfilm I happened to possess of Ludovico Carbone's *Additamenta* to the logic textbook of Franciscus Toletus (Venice 1597). To my surprise I found a number of word-for-word parallels between the latter work and Galileo's manuscript. This served to confirm a thesis to which I was inclining at the time, which had been reinforced by Crombie's earlier discovery that two questions in MS Gal. 46 seem to have been extracted from Christopher Clavius's *Sphaera* (Rome 1581). This was that Galileo's early manuscripts were directly connected with the teachings of Jesuits at the Collegio Romano. But in view of the facts that the transcription of MS Gal. 27 was for my private use only, and that Crombie and Carugo had assured me that it was to be published within a year, I simply filed it away and continued to focus my attention on the sources of MS Gal. 46.

In 1975 I was back in Italy again, this time to investigate handwritten *reportationes* of lectures given at the Collegio Romano as possible additional sources of MS Gal. 46. In effect the only printed sources of late sixteenth-century Jesuit teachings relating to that manuscript, apart from Clavius, were the textbooks of Franciscus Toletus and Benedictus Pererius, and a thorough search of all three had turned up only fifteen percent of the materials discussed in the manuscript. In the intervening period, however, I had obtained microfilms of Pererius's lectures at the Collegio from the Oesterreichische Nationalbibliothek in Vienna and found that they covered matter very similar to that of Galileo's "Physical Questions." Through the good graces of Edmund Lamalle and Vincenzo Monachino, librarians respectively of the Roman Archives of the Society of Jesus and of the Archives of the Pontifical Gregorian University, in June of 1975 I obtained access to scores of manuscript teaching notes of lectures on physics that had been given at the Collegio from the late 1570's onward. A search through these enabled me to identify rather quickly parallel texts for an additional seventy-five percent of the matter contained in MS Gal. 46. These new texts supplied me with all the information I needed to complete my commentary on that manuscript, and I proceeded with the publication of *Galileo's Early Notebooks: The Physical Questions*, which came out, as already mentioned, in 1977.

In the course of my investigations in 1975 relating to MS Gal. 46 I became aware of the teachings of a Roman Jesuit, Paulus Vallius, whose physical questions, though only partially available, manifested more affinities with Galileo's manuscript than did those of other Jesuits. A search that year through the rare book collection of the Biblioteca Nazionale in Rome then turned up a copy of Vallius's two-volume *Logica* (Lyons 1622). Perusal of the contents of these volumes alerted me to similarities between them and Ludovico Carbone's *Additamenta*, both of which are described in the Introduction that follows. We now know, as a matter of fact, that Carbone plagiarized the logic course Vallius had given at the Collegio Romano in 1587–1588, that somehow Galileo himself gained access to the same course, and that a portion of it dealing with Aristotle's *Posterior Analytics* actually served as the exemplar for MS Gal. 27. In 1975 none of these connections had yet dawned on me or anyone else. In fact it was not until 1980 – when, having obtained microfilms of Vallius's *Logica*, I read through the prefaces to both volumes – that I became aware of Vallius's charges of plagiarism directed against Carbone. This supplied me with a much-needed key to the provenance of MS Gal. 27, and also of MS Gal. 46, for I then suspected that there were affinities between the two manuscripts, and, moreover, that the sources of MS Gal. 27 might be easier to identify than those of MS Gal. 46. This realization reawakened my interest in MS Gal. 27, for what obviously was required at that point was a detailed exploration of the connections between its teachings and those of Carbone and Vallius.

During my travels in Italy in 1975 I had encountered Carugo in Milan and had told him about the parallels between Carbone's *Additamenta* and MS Gal. 27, which he himself had discovered in the meantime, and also of the similar materials in Vallius's *Logica,* of which he knew nothing, and of which I then took the opportunity to inform Crombie by mail. At the time I was not aware that Carbone had plagiarized Vallius's lecture notes, not having discovered this myself until May of 1980, but neither Crombie nor Carugo manifested any interest in the clues about Vallius I was providing them. By 1980, moreover, they had still not published Carugo's long-promised transcription of MS Gal. 27. Thus, when I was alerted to the probability of Carbone's plagiarism, I felt stymied by not being able to use Carugo's work or, alternatively, to continue my transcription of MS Gal. 27 when it was known that Carugo had provided me with a copy of his. I discussed this situation with the late Charles B. Schmitt, who had studied with Edwards at Columbia University under Paul Oskar Kristeller. Schmitt informed me that Edwards had already transcribed the Galileo manuscript, in complete independence of Carugo and Crombie, and that he might be willing to share his transcription with me. At Schmitt's instigation I contacted Edwards, who had not worked on the manuscript for ten years, but was interested in the new discoveries and was happy to entrust his transcription to me. This, it turned out, was more meticulous and detailed than Carugo's and was ideal for my purposes. Using it and sources I had identified in the interim, I ultimately was able to date the origin of MS Gal. 27 as well as of MS Gal. 46 and to pinpoint the exemplars from which they both derived. I published these findings in my *Galileo and His Sources: The Heritage of the Collegio Romano in Galileo's Science* (Princeton 1984).

The idea of a collaborative effort between Edwards and myself to produce the Latin Edition of MS Gal. 27 was first broached at an International Congress on "Aristotelismo Veneto e Scienza Moderna," held at the University of Padua in September 1981 under the auspices of the Centro per Storia della Tradizione Aristotelica nel Veneto. Aware of the renewed interest in Galileo's logic and its possible relationship to that of the Paduan Aristotelians, and alerted to questionable interpretations of MS Gal. 27 (as well as MS Gal. 46) that were voiced by Carugo at the Congress, Edwards, Schmitt, and myself – all of whom were present there – discussed among ourselves the possibility of its publication. The specific idea that emerged was to have Edwards rework his transcription of the Latin text while I was preparing an introduction, notes, and commentary in English, then to have the volume put out by the Centro as part of its publication series. We outlined this proposal to Ezio Riondato, Director of the Centro, and received his enthusiastic support. I then returned to the University of Padua for two months during the academic year 1983–1984 to lecture on MS Gal. 27 and its sources. During that period I worked out the problem with the demonstrative *regressus* that had bothered Edwards (*see* Lat. Ed., 288–302), and mapped out the work that remained to be done. The completed edition, entitled Galileo Galilei, *Tractatio de praecognitionibus et praecognitis* and *Tractatio de demonstratione*, was published by Editrice Antenore in 1988, with Edwards being responsible for the transcription and its apparatus and I for the introduction, notes, commentary, bibliographies, and indexes.

In view of the book's being published in Italy there was no thought at the time of including in it an English translation of the Latin text. All agreed, however, that this would be a desideratum in view of the fact that much research on Galileo's logical methodology had been going on in the U.S. and the U.K. Accordingly, aided by a grant from the National Endowment for the Humanities, I set to work on making Galileo's logic intelligible to the Anglo-American world. The present volume is the fruit of that enterprise.

In view of the needs of most English readers the translation of the manuscript presented herein is as literal as possible consonant with clear expression, with technical terms given in vernacular equivalents and the Latin supplied in notes. The text translated is obviously not a finished piece of work but represents the effort of a young professor to abbreviate and reformulate teachings contained in a classic of logic, difficult enough in the original Greek and no less so in Latin versions. On this account Galileo's style is more cryptic than one might expect, and there are occasional slips or errors in his exposition. In some cases interpolations had to be made in his text to complete a thought or simply make it intelligible to modern readers; in others, corrections or emendations had to be introduced in order to render his thought consistent. Insertions of the first type I have shown in square brackets [], those of the second in curly brackets or braces {}. The rationale behind the latter I explain in each case in an accompanying note.

In place of the line numbers used in the Latin Edition, I have numbered the paragraphs of Galileo's text in that edition successively, and have used these numbers in this volume for cross-references. As will be seen, the manuscript contains two treatises, the first dealing with the foreknowledge required for demonstration (designated F), the second dealing with demonstration itself (designated D). Each treatise contains three disputations, each disputation is divided into questions, and each question is divided into paragraphs. A sequence of numbers can thus be used to designate unambiguously each paragraph in the manuscript. Thus F3.2.4 refers to the treatise on foreknowledge, third disputation, second question, paragraph 4; D2.6.9 to the treatise on demonstration, second disputation, sixth question, paragraph 9. For the benefit of the reader who wishes to use the translation in conjunction with the Latin Edition or Galileo's autograph, I have included a concordance at the end of the volume; this lists, for each paragraph of the translation, the page and line number at which it begins in the Latin Edition, and the folio and line number at which it begins in MS Gal. 27.

It should be noted that the commentary on the translation is different in most particulars from the notes and commentary in the Latin Edition. The latter are largely paleographical and source oriented, their aim being to provide evidence of copying or of derivation from the exemplar used by Galileo in writing the manuscript. Where relevant, such sources are there cited in Latin to enable the reader to identify parallel passages in Galileo's appropriation. Obviously this type of annotation takes for granted not only a knowledge of Latin but a more than passing acquaintance with medieval and Renaissance commentaries on the *Posterior Analytics*. The comments in this volume have a different purpose altogether and make no such suppositions. They serve rather to introduce the reader to the technical language of the *Posterior Analytics*, to signal other parts of Galileo's text that are presupposed to, or further elaborate on, the passage being commented on, and to make reference to the background out of which the text emerged or to more systematic elaborations of its contents that may be found in other works.

In view of the fact that the sources and dating of MS Gal. 27 are closely connected with those of MS Gal. 46, and also of a third manuscript (MS Gal. 71, which contains Galileo's early treatises on motion) – all of which were probably written at Pisa between 1588 and 1591 – I have prefaced my translation and commentary with an Introduction that describes rather fully Galileo's early Latin manuscripts. Its purpose is not only to bring my introduction to the Latin Edition of MS Gal. 27, published in 1988, up to date, but also to supplement the documentary evidence provided in *Galileo and His Sources*, published in 1984, as well as that presented in *Galileo's Early Notebooks*, published in 1977. Here the reader will find the most recent information on the dating and likely sources of all three of Galileo's Pisan manuscripts, as well as a more detailed analysis of the sources of MS Gal. 27. The concluding section of the Introduction then provides an all too brief evaluation of how Galileo's appropriated treatises on Aristotle's *Posterior Analytics* may have influenced the development of his science. For a fuller account the reader should consult *Galileo's Logic of Discovery and Proof*, for reasons now to be explained.

As will become clear from the Introduction, the treatises found in MS Gal. 27 were appropriated from but a small portion of a logic course being taught at the Collegio Romano in the last two decades of the sixteenth century. The course itself, which has been described in general terms in *Galileo and His Sources*, lasted for an entire year and covered all the materials found in Aristotle's *Organon*. The part of the course Galileo copied out, namely, that devoted to the first book of the *Posterior Analytics*, was taught in about a month and a half toward the end of the course. The text of the *Analytics* is itself difficult to understand, and indeed most of the prior instruction in the course was directed to building up the knowledge required for its comprehension. To further complicate matters, the concepts of logic elaborated previous to the part of the course Galileo appropriated and the concepts of science treated in the part following it are very different from those taught in logic courses in the present day. Yet they are crucial for understanding not only the *logica docens* (or "logic teaching") contained in MS Gal. 27 but also the *logica utens* (or "logic in use") Galileo employed in his various scientific investigations.

Awareness of this situation, which became progressively clearer as I worked through the translation and the materials required for its proper comprehension, led to another fruit of the enterprise sponsored by the National Endowment. This is a companion volume, entitled *Galileo's Logic of Discovery and Proof*, much of which was written subsequent to my preparing the translation and commentary. Actually it has turned out to be both propadeutic and complementary to the translation, since it lays out in full detail the logical system Galileo appropriated and then explains how he employed it in developing his new sciences of the heavens and of local motion. The translation volume, of course, is still basic, since it documents Galileo's actual wording of the logical teaching he appropriated from the Jesuits. But to make it more useful to the reader who desires a deeper understanding of the epistemology and ontology that lies behind that teaching, I have cross-referenced the commentary of this volume to the various sections into which Galileo's *Logic of Discovery and Proof* is divided. Furthermore, at the beginning of the notes for each of the questions translated, I have indicated the sections that should be read both for background to the teaching it contains and for applications in Galileo's subsequent writings. The companion volume is thus the major source to which one should turn for a fuller comprehension of the manuscript and the way in which its logic guided Galileo in elaborating the "new sciences" for which he is justly famous.

I should like to thank the National Endowment for the Humanities, an independent agency of the U.S. government, for financial support through its Research Grant RL–21080–87, without which this project would never have been begun. I also wish to acknowledge the invaluable assistance of William F. Edwards, whose painstaking and accurate work in transcribing MS Gal. 27 made the Latin Edition possible. Other scholars to whom I am indebted include the late Charles B. Schmitt for encouraging Edwards and me to begin the project, though he did not live to see its completion; the late Hippocrates G. Apostle for his accurate and informed translations and commentaries on Aristotle's text; Mario Otto Helbing, formerly of the Scuola Normale Superiore of the University of Pisa, for his exhaustive research on Francesco Buonamici; the faculty of the Centro per la Storia della Tradizione Aristotelica nel Veneto of the University of Padua, especially Ezio Riondato, Enrico Berti, Antonino Poppi, and Luigi Olivieri, for their encouragement and collaboration at all stages of the work; and my friend and colleague, Jean Dietz Moss, who has been a benefactor to me in countless ways throughout the entire enterprise. I further owe an expression of gratitude to Everard de Jong and other graduate students at The Catholic University of America who followed with much enthusiasm my course on Aristotle's *Posterior Analytics* over a period of two decades. Finally, I am deeply grateful to Professor Robert S. Cohen of Boston University, editor of the Boston Studies in the Philosophy of Science, for publishing the results of my work in his distinguished series.

W.A.W.
College Park, Maryland

ABBREVIATIONS

F	Treatise on Foreknowledges and Foreknowns
D	Treatise on Demonstration
[3]	Paragraph 3
[3a]	Subdivision of paragraph 3
F3.2.1	Treatise on Foreknowledges and Foreknowns, Disputation 3, Question 2, Paragraph 1
F3.2 n. 2	Note 2 to Disputation 3, Question 2
71a12	Reference to the Greek text of Aristotle prepared by I. Bekker (Berlin 1831–1870), page 71, column a, line 12, hence referred to as a Bekker number
text 2	Reference to a paragraph or portion of the text of Aristotle, identified by Bekker numbers in the notes
Sec. 4.3d	Section of *Galileo's Logic of Discovery and Proof*, Chapter 4, Section 3, Subsection d
[...]	Inserted into text by translator, because either omitted in Galileo's manuscript or required for sense
{...}	Changed in text by translator, because of an error or misreading in Galileo's manuscript
Lat.	The Latin for the word or passage commented on
Gr.	The Greek for the word or passage commented on
Lat. Ed.	Galileo Galilei, *Tractatio de praecognitionibus et praecognitis* and *Tractatio de demonstratione*, eds. W. F. Edwards and W. A. Wallace. Padua: Editrice Antenore, 1988.

SOURCES: SHORT TITLES

CA	Carbone, *Additamenta*, Venice 1597
GG	Galileo, *Le opere di Galileo Galilei*, ed. A. Favaro, 20 vols. in 21, Florence 1890–1909. repr. 1968
LL	Lorinus, *Logica*, Cologne 1620
VL1	Vallius, *Logica*, Vol. 1, Lyons 1622
VL2	Vallius, *Logica*, Vol. 2, Lyons 1622
Vallius-Carbone	Attribution of dual authorship for lecture notes of Vallius, subsequently published by Carbone

INTRODUCTION

INTRODUCTION

It is now generally agreed that Galileo composed three Latin manuscripts while he was teaching, or preparing to teach, at the University of Pisa toward the end of the sixteenth century, in the years roughly between 1589 and 1591. One of these, MS 71, contains his earliest attempts at constructing a science of motion. It gives evidence throughout of being an original composition, revised and even recopied in places, but all of it written by Galileo in his own hand. MSS 27 and 46, which contain his logical and physical questions respectively, are also autographs, but they show numerous signs of copying. Though this fact was recognized by Antonio Favaro, the editor of the National Edition of Galileo's works, at the end of the nineteenth century, he speculated then that both could only be student compositions. The first he assigned to the period of Galileo's studies at the Monastery of Vallombrosa in the late 1570's [GG9: 279]; the second he thought must have been written while Galileo was a student at the University of Pisa, dating it in 1584 and pointing to Francesco Buonamici, one of Galileo's professors there, as its likely source [GG1: 12]. Favaro provided no direct evidence in support of either of these identifications, but there was no reason to question them and they were generally accepted among scholars for close to a century.

Recent studies by Adriano Carugo, Alistair Crombie, Christopher Lewis, William F. Edwards, and the translator have now raised doubts about such early datings and the sources they would seem to entail.[1] The sophistication and the general erudition of the notes in MSS Gal. 27 and 46 argue against their being composed, or even excerpted from other sources, by a mere student, however precocious he might be. It has also been possible to identify specific texts on which both manuscripts could have been based. In light of these discoveries a wholesale reassessment of the manuscripts and their importance for Galileo's intellectual development has recently been in process. Among the problems being addressed is whether Galileo used printed sources when preparing these notes or whether he based his compositions on manuscripts or other hand-written sources, and, if so, how he obtained them and what his motivation might have been for doing the considerable work involved in making

3

copies for himself. Before discussing these questions, however, it will be well to give a description of the manuscript that is the principal focus of interest in what follows, MS 27, and a brief overview of its contents.

DESCRIPTION OF MS 27

The codex containing Galileo's appropriated commentary on Aristotle's *Posterior Analytics*, numbered 27 in the collection of Galileo's writings now preserved in the Biblioteca Nazionale Centrale in Florence, is described by Angiolo Procissi in the catalogue of that collection.[2] It is written on paper of various types, which Procissi dates from the seventeenth century, and is composed of two elements that differ in form. The first group, made up of folios 3 to 32 inclusive, are 217 × 150 mm. in size, and the second, composed of folios 34 through 46, are 240 × 188 mm. in size; they are separated by a white sheet of modern paper, numbered folio 33. At the beginning there is a gray fly-leaf, neither numbered nor written upon, followed by two modern sheets, numbered 1 and 2, on the first of which is inscribed, on its front side, the title, *Opere / di / Galileo Galilei / Parte I / Tomo 17 / Dialettica*, and on the second, again on the front side, the index to the volume. Following these two sheets the remaining folios are numbered sequentially. At its end there are again two blank sheets, white in color and numbered 47 and 48, followed by a gray fly-leaf, neither numbered nor written upon. In all this makes up 48 folios, not counting the fly-leaves, and these are numbered in pencil on the upper right of their front sides in a modern hand. Seventeen of the total 96 sides are blank, namely, 1v, 2v, 32r–v, 33r–v, 43r–v, 44r–v, 45r, 46r–v, 47r–v, and 48r–v. The codex is bound in cardboard covered with maroon leather. On the spine is printed in gold, on a black label, the title: *Galileo, Dialettica, P.I.T.XVII.*

In this codex the logical treatises occupy folios 4r through 31v. They are from a small notebook, mutilated at the beginning, and written in Galileo's hand. Procissi identifies them as a scholastic exposition of various questions in "dialectics," a term generally applied to the whole of logic in Galileo's day. In Favaro's judgment, as already noted, Galileo merely played the role of an amanuensis in their composition; for that reason Favaro transcribed only one question and then provided a listing of the various treatises and questions in the portion of the notebook devoted to logic [GG9: 273–292]. Procissi remarks that the first disputation is missing from the treatises as a result of the mutilation. In addition, Galileo makes reference on folio 13r to a following treatise on science, and this is not in the notebook; since his cross-references throughout its surviving portions are generally accurate, this could be an indication of materials missing at the end of the notes as well as at the beginning.

The titles of the logical treatises and of the questions they contain is given in English translation in the Table of Contents. (The Latin is to be found in the Latin Edition, which gives all available details relating to Galileo's Latinity.) In the manuscript the Treatise on Foreknowledges and Foreknowns begins on folio 4r and concludes on folio 13r; the Treatise on Demonstration begins directly under this on the same folio and concludes on folio 31v, about halfway down the page, the rest of which is blank. The folio and line numbers at which individual questions begin are given below in Table 8 and in the Concordance at the end of the volume. (If a Treatise on Science were present, it would have begun at the top of folio 32r in the original foliation.) The questions now extant are 28 in number, 11 pertaining to foreknowledge and 17 to demonstration, and there are 300 paragraphs in all, 95 in the treatise on foreknowledge and 205 in that on demonstration. All of the questions are relatively short, D1.2 being the longest with 37 paragraphs and D3.2 with three, the average being about eleven. Galileo's handwriting is generally larger at the beginning of the codex than at the end, with the result that an average of only 24 lines per page are found on the first few pages and as many as 47 lines on the last. There are many signs of copying in the manuscript, and many errors in Latinity; these are not enumerated in what follows since they are adequately discussed in the Latin Edition.

Following the logical treatises, the remainder of the codex contains excerpts from the moral writings of Plutarch (folios 34r through 42v) and a rendering in verse of some of the teaching contained therein (folio 45v). It is likewise an autograph, and is made up of parts taken from two different notebooks, corresponding to folios 34 through 38 and 39 through 46 respectively. The titles and locations of these excerpts are given by Procissi. There he also notes that Favaro has identified their sources in two printed works: the earlier ones and the last are taken from a translation of Plutarch's *Opuscoli Morali* published at Venice in 1559, whereas the later ones (the last excepted) is from Lodovico Domenichi's translation of the same, published at Lucca in 1560.[3]

The codex has been examined for watermarks by Adriano Carugo and Alistair Crombie, and independently by Stillman Drake.[4] Carugo and

Crombie report that the paper on which the logical treatises were written is unwatermarked, whereas Drake sees evidence on some folios of a bell-shaped watermark that is uncircled, which he conjectures to be of Pisan origin. Drake's dating of the Pisan manuscripts, MS 27 included, made on the basis of these watermarks generally agrees with our own, to be discussed below. Carugo and Crombie, on the other hand, propose very different datings on the basis of published works from which they think two of the manuscripts in which we are interested, MSS 27 and 46, are derived. Their datings raise the question whether Galileo's sources were actually published works, as they surmise, or manuscripts, as held first by us and now also by Drake.[5]

GALILEO'S SOURCES: MANUSCRIPTS OR PRINTED WORKS?

From internal evidence, to be discussed later, it seems clear that MS 27 was composed before MS 46. The latter had been transcribed and published by Favaro in the National Edition, however, and this perhaps explains why its Latin text was the first to be studied. After considerable investigation it was found to contain a number of textual parallels with four textbooks published by Jesuit professors at the Collegio Romano. These are described in the literature and may be enumerated as follows: (1) a *Physics* commentary and questionary by Franciscus Toletus, printed at Venice in 1573; (2) a *De generatione* commentary and questionary by the same author, printed at Venice in 1575; (3) a work on natural philosophy by Benedictus Pererius, printed at Rome in 1576; and (4) a commentary on the *Sphere* of Sacrobosco by Christopher Clavius, printed at Rome in 1581. All of these dates are prior to 1584, and thus the discovery of these parallels between Galileo's manuscripts and the four printed works does not invalidate Favaro's conjecture that MS 46 was composed while Galileo was a student at the University of Pisa in 1584. The parallels, however, account for at most 15 percent of Galileo's entire composition in that manuscript, and they leave open the question of other sources on which its remainder might have been based.

As mentioned in the preface, two transcriptions of the Latin text of MS 27 have been made independently, one by Edwards and the other by Carugo; their transcribed texts reveal further connections with the Collegio Romano. Galileo's logical questions show decided similarities with Toletus's *Logica*, a textbook published in 1576, only a year or two before Favaro's conjectured date for the writing of MS 27. More surprisingly, however, they show a large number of textual parallels with an *Additamenta* to Toletus's *Logica* that was put into print by a certain Ludovico Carbone, who published this at Venice, but not until 1597. Now if Galileo used Carbone's *Additamenta* to compose his logical questions, he would have been at least thirty-three years old, at the height of his teaching career at the University of Padua. Moreover, since the logical questions quite clearly antedate the physical questions, this would mean that both of the manuscripts in question, MSS 27 and 46, could not have been written by Galileo until after 1597, close to twenty years later than Favaro first speculated for their composition. A yet further complication comes from the fact that some memoranda on motion written at the back of MS 46 form the basis for several treatises on motion that are found in the third Pisan manuscript, MS 71, which scholars agree was composed between 1590 and 1592, while Galileo was teaching mathematics at the University of Pisa. Thus to claim that MSS 27 and 46 were based exclusively on printed sources presents a number of difficulties. MSS 46 and 71 could have been composed prior to 1592, at which time Galileo moved from Pisa to Padua, but MS 27, which gives clear indication of having been written before the other two, could not have been written until 1597, and this would be a full five years after the move.

Such difficulties notwithstanding, Carugo and Crombie have argued that the main source of MS 27 was actually Carbone's *Additamenta*, and that the consequent late dating of that manuscript necessitates a wholesale revision of the accepted chronology of Galileo's writings and the place of his scholastic compositions within them. Rather than see MSS 27 and 46 as student lecture notes, the authors contend – from their later dating and the mature style of arguing in the manuscripts – that they provide "a systematic and learned account of a series of philosophical questions and disputations compiled by an experienced scholar whose interests are focused on the theory of science and demonstrative knowledge and on cosmology and natural philosophy."[6] Such a stance requires them to find a new date for the composition of the treatises on motion contained in MS 71. These they assign various dates: the earliest among them they admit could come from Galileo's Pisan days, but the later versions they hold were not completed until after his *Meccaniche*, which they would date sometime between 1615 and 1623. The treatises on cosmology and natural philosophy contained in MS 46 were then compiled, on their reckoning, either while writing or just after publishing *Il Saggiatore*, which they point to as the first published work in which Galileo indulges in scholastic disputations. This new estimate of notes Favaro had labeled *Juvenilia* puts them in reasonable proximity to the writing of the *Two World Systems*, many of whose discussions, they argue, can be traced to scholastic commentaries circulated or produced by Jesuits between the end of the 16th and the beginning of the 17th century. So Carugo and Crombie see the famous *Dialogo* of 1632 as actually based on two models: those of Plato's dialogues and of the scholastic disputation revived by the Jesuits and practiced by Galileo himself in his scholastic dissertations.

This obviously is revisionism of a drastic sort, and it has had little or no acceptance among Galileo scholars. We thus consider the alternative and return now to the dating of the three manuscripts we have proposed in previous publications, which sees the manuscripts as themselves based on manuscript sources and not on works already in print.

Focusing first on MS 46 and the account given in *Galileo's Early Notebooks*, as early as 1972 we had begun to wonder why only 15 percent of that manuscript shows parallels with the printed texts of Toletus, Pererius, and Clavius mentioned above. Our study of corrections, deletions, and other signs of copying in the manuscript raised suspicions that many of these were occasioned by Galileo's working from handwritten materials he had difficulty deciphering. A check into *reportationes* of Pererius's lectures on natural philosophy conserved in the Oesterreichische Nationalbibliothek soon confirmed that suspicion: the lectures contained more material than was printed in his textbook, and much of the new material had counterparts in Galileo's manuscript. Further checking into other *reportationes* of lectures given at the Collegio, lectures that never did find their way into print, disclosed that an additional 75 percent of Galileo's composition had correspondences in Jesuit teaching notes. These new parallels, some of which are reproduced in our *Prelude to Galileo* and others in our *Galileo and His Sources*, also served another purpose: they provided strong evidence for dating Galileo's writing of MS 46 around 1590. The closest textual parallels, in fact, are with the notes of a Jesuit professor, Paulus Vallius, who taught the course in natural philosophy at the Collegio in 1588–1589 and whose notes were not available until late in 1589. This later dating, some five years beyond that proposed by Favaro, fits in well with the previously noted connection between MS 46 and MS 71, since it confirms that both were written while Galileo was teaching at the University of Pisa. Thus the first is not a student composition, as Favaro thought, and though written at Pisa, not in the way he had conjectured.

VALLIUS AND CARBONE

That still leaves the problem of dating MS 27. Fortunately there is a *rotulus* of professors who taught in the Jesuit college at Rome in various years, and this reveals that Vallius had the logic course in 1587–1588, that is, the year immediately preceding his teaching the materials appropriated in MS 46. Not only this, but the handwritten list indicates that a professor named "Ioannes Lacerino" taught logic at the Collegio two years before Vallius, in 1585–1586. Our subsequent investigation turned up a Latin manuscript in the Vatican Library with logic notes ascribed to a certain "Ioannes Laurinis S.I." and noting that the latter had taught logic in Rome in 1584. In the family name recorded on the manuscript the letters "au" had been crossed out and the letter "o" written above them, thus correcting the author's name to Lorini, or, in Latin, Lorinus. Now Lorinus, it turns out, developed into a Scripture scholar and later produced a number of publications, among which is a textbook on logic printed in 1620. When that textbook was located and its contents compared with the manuscript version in the Vatican Library, the result was surprising. The published course was almost exactly the same as that existing in manuscript some 36 years previously!

Vallius's Logica. Apparently this was not an isolated instance, for a search through the card files of Italian libraries revealed that Vallius had also published an entire course in logic based on the lectures he had given in 1587–1588. This appeared in two massive folio volumes entitled *Logica*, printed at Lyons, but not until 1622. Such a late date was somewhat discouraging, and so was the material contained in the tomes. Somewhat like Lorinus's text, yet in far more exhaustive detail, they covered the same ground as did Galileo's MS 27, but not in precisely the same words. Repeated checking made it clear that there were no parallels in Vallius's printed text superior to those already identified in Carbone's *Additamenta* of 1597, a work published 25 years previously and already too late for fixing the date of Galileo's writing MS 27 in his early years at Pisa.

This perplexing situation was finally resolved through our reading of the prefaces of the two volumes of Vallius's *Logica*. In his remarks "To the Reader" at the beginning of Volume I, after first outlining the contents of the volume, Vallius explains that he will start by prefacing a brief introduction to the whole of logic. Of this he writes:

We preface, I say, an *Introduction* that was explained by us 34 years ago [i.e., in 1588] in the Collegio Romano and given to our hearers shortly thereafter. This work, with very little of the fruits of our labors changed in it, was published at Venice by some good author, who added some preliminary matter and made some inversions (or rather perversions) of its order that, in my judgment, achieve no better results. We wish to warn you the reader of this, so that, should you come across this book, you will recall that he took it from us. And since he stole this and similar matter from us and from the writings of our Fathers [i.e., other Jesuits], perhaps he should have added the author's name to these books, had he known it or thought it due us [VL1: 4].

Vallius then announces that his second volume will contain his expositions of the *Prior Analytics* and the *Posterior Analytics*, and to this he will add a disputation on science, of which he further remarks:

The same thing happened to this *Disputation* as I explained happened to the *Introduction*. But this we have now so enlarged and perfected that it would hardly be recognized by anyone except the author as the fetus of the same [VL1: 4].

Although Vallius does not name the "good author" who plagiarized his work, there can be no doubt that this was Ludovico Carbone, who in 1597 published at Venice his *Introductio in logicam*, which contains the *Introduction* to which Vallius refers, and then in 1599, also at Venice, printed an *Introductio in universam philosophiam*, which includes the *Disputation* on science here described by Vallius.

The preface to Vallius's second volume is even more helpful for establishing Carbone's identity. By the time he had come to its preparation Vallius had decided that he would append four complete tractates to his commentaries on the *Analytics* and not merely the one disputation on science. These he now enumerates as treatises on foreknowledge, on demonstration, on definition, and on science. Concerning the order of these treatises he alerts the reader to the following:

About twenty years ago [i.e., around 1600], a certain individual – possessing a doctorate, having published a number of small books, and being otherwise well known – had a book printed at Venice in which he took over and brought out under his own name a good part of what we had composed in our *On science* and had taught at one time, 34 years before this date [i.e., in 1588], in the Roman *gymnasio*. And having done this, this good man thought so much of other matters we had covered in our lectures that he took from them, and claimed under his own name, a large part of *On the syllogism*, *On reduction*, *On foreknowledges*, and *On instruments of scientific knowing*, and proposed these as kinds of *Additamenta* to the logic of Toletus, especially to the books of the *Prior Analytics*. He also saw fit to publish, again under his own name, our *Introduction* to the whole of logic, having changed only the ordering (disordering it, in my judgment), along with the introductions and conclusions. I wish you to know this, my reader, so that, should you see anything in either, you will recognize the author. I say, "should you see anything in either," for we have so expanded our entire composition that, if you except only the opinions (which, once explained, we have not changed), hardly anything similar can you see in either. So in those works you have what he took from me, in this what I have prepared more fully and at length [VL2: 1].

Here, then, was the solution to the puzzle. The "good man" to whom Vallius makes reference was no doubt Ludovico Carbone, for Carbone did publish a number of small books (*libelli*) between 1587 and 1597.[7] Indeed, in 1588 he put out an edition of Toletus's introduction to dialectics (*Introductio in dialecticam*), acknowledging in it that Toletus was his teacher, and adding to it some preludes (*praeludia*) and tables composed by himself. In its preface he also mentions that he taught an introduction to logic in Rome in that same year (1588), to the great benefit of his students. In 1597, moreover, he published at Venice the *Additamenta* to which Vallius refers, whose full title translates as "Additions to the Commentaries of Franciscus Toletus on the Logic of Aristotle: Preludes to the Books of the *Prior Analytics*; Treatises on the Syllogism, on Instruments of Scientific Knowing, and on Foreknowledges and Foreknowns." These are precisely the disputations Vallius charges Carbone with appropriating, and the conjunction of their titles with an infrequent collective noun such as *Additamenta* places beyond doubt the identity of the "good man" who put them together. Also to be noted is Vallius's statement that he did not change his explanations of the opinions and positions on the various questions. We may reasonably expect from this that the citations of authorities and sources in the printed text of 1622 will be the same as in the lecture notes of 1588.

On the basis of this evidence, the materials contained in Carbone's *Additamenta* are considered throughout this volume to be the product of joint authorship. They are therefore attributed to Vallius-Carbone and assigned the dates 1588–1597, the first being the date when Vallius's exemplar was completed and the second that of Carbone's publication. A few details about Carbone and his activities may now prove helpful for forming a judgment of the respective contributions of the two authors to the final result.

Carbone's Work.   Much of what is known about Ludovico Carbone is the result of research by Jean Dietz Moss, who has been interested in the history of rhetoric in the Renaissance and the role played in that history by the Jesuits of the Collegio Romano.[8] She speculates that Carbone was a student with them in the early 1560's, where he was a devoted member of one of their sodalities, the Congregatio Beatae Mariae Annuntiatae. This squares with Carbone's statement that he had studied logic under Toletus, for Toletus taught the logic course at the Collegio in 1559–1560. Early in his career he prepared a study guide for a popular compendium of rhetoric by the Portuguese Jesuit, Cipriano Soarez, entitled *De arte rhetorica*, which Carbone claims to have seen in manuscript. This was first published in 1562 and then reprinted 134 times down to 1735. In the preface to this guide, which bore the title *Tabulae Rhetoricae Cipriani Soarii* and which he published along with his commentary, *De arte dicendi*, in 1589, Carbone praises his professors at the Collegio for the content of their lectures and the manner in which they instructed their students. He wrote then that he was actively preparing ten more works that would benefit not only his own students but also those who attend Jesuit schools. Moss suspects that some of these are based on *reportationes* of lectures on rhetoric she has uncovered at the Collegio Romano. Since these are of Jesuit authorship, this would corroborate Vallius's charge that Carbone had stolen not only his writings but also those of other Jesuits. She has also discovered a manuscript containing Carbone's commentaries on the *De caelo* and *De generatione* which bear considerable affinity to Jesuit lectures on these works and also to Galileo's MS 46.[9] Yet her judgment of Carbone's appropriation of all these materials is less harsh than one might expect, for she concludes on the note that "the piety he exhibits towards his mentors and his own expressed concern to further their teachings helps to excuse his lack of concern for what we moderns would term plagiarism."[10]

Summing up Moss's and other evidence with regard to Carbone, we may gather that: (1) he had studied philosophy at the Collegio prior to 1562, probably beginning in 1559 under Toletus; (2) he taught an introduction to logic in Rome in 1588 and probably at the Collegio, where Vallius had taught a similar introduction in late 1587; (3) he published in 1588 some preludes to, and tables for, Toletus's introduction to logic, adumbrating materials that would later appear in his own *Introductio* of 1597, now known to be based on Vallius's lectures; (4) he published in 1589 a guide and tables to Soarez's rhetoric, at which time he stated that he was preparing ten more works related in some way to Jesuit teachings; (5) he wrote out in 1594 manuscript notes on the *De caelo* and *De generatione* based on Jesuit materials, probably also in Rome and possibly one of the ten works then under preparation; (6) he published in 1597 and under his own name an *Introductio in logicam*, appropriating Vallius's similar introduction from the latter's lectures given in 1587; (7) in 1597 he likewise published under his own name an *Additamenta* to Toletus's *Logic*, similarly appropriating materials in it from Vallius's logic course; and (8) in 1599 an *Introductio in universam philosophiam* was published under his name (posthumously), containing a treatise *De scientia* alleged by Vallius to be also based on the latter's logic course.

On the matter of Carbone's possible contribution to all of these reworked treatises, Moss's considered view is that his part was mainly that of the pedagogue – ordering the materials so that their connections could readily be seen and supplying apt illustrations to make them intelligible. In her judgment he was an outstanding teacher. His style, she writes, "is direct but elegant, not flowery but explicit, and replete with pertinent illustrations from the classics."[11] Apart from Vallius's prefaces, therefore, on the bases of Carbone's own testimony and Moss's study of his style, it seems reasonable to attribute the basic content of the *Additamenta* and the appropriated treatises in the *Introductio in logicam* and the *Introductio in universam philosophiam* to Vallius's lectures of 1587–1588. The stylistic and pedagogical innovations in them, on the other hand, would seem to be the result of Carbone's editorial work in 1597 for the *Additamenta* and the *Introductio in logicam*, and some time later for the *Introductio in universam philosophiam*.

Dating the Manuscripts. The dating problem posed by Carugo and Crombie yields quickly to solution on the basis of this same evidence. Following the traditional view, MSS 27, 46, and 71 were composed in the order of their enumeration, beginning probably in 1589 (since Vallius did not complete his logic course until August of 1588 and did not give out his notes until "shortly thereafter"), and finishing equally probably in 1592, before Galileo's move to Padua. This places the writing of all three manuscripts either contemporaneous with, or immediately preparatory to, his teaching career at the University of Pisa. Not only was MS 27 based on a handwritten *reportatio*, but so was MS 46, although the *reportationes* on which it was based could have been excerpted from the printed texts of Toletus, Pererius, and Clavius – which might explain the fifteen percent of textual parallels with those works in Galileo's manuscript.

This, then, is the answer to the question posed in the title to the preceding section, namely, whether Galileo's sources were manuscripts or printed works. His primary sources were handwritten, although portions of them may have incorporated materials that had already found their way into print. When we take the manuscript evidence into account, therefore, we need not subscribe to the chronology proposed by Carugo and Crombie, while we can still account for many of the scholastic influences to which they call attention and which are manifest in Galileo's later writings. In this way the established dating of the remainder of Galileo's works is preserved, and yet the heritage of the Collegio Romano is seen as an early and important factor shaping the overall development of Galileo's science.

POSSIBLE SOURCES OF MS 27

Much of the foregoing research would not have been possible had not a copy of the *rotulus* of professors at the Jesuits' Roman college, indicating their subjects and the years in which they taught, survived to the present. Apparently it was a custom for each professor to deposit a set of his lectures in the Collegio's library. Some of these were sent to other institutions, usually Jesuit, to serve as models there, and yet others were copied and recopied for various purposes. Only a small number of these are extant, but fortunately enough of them are available from the period around 1590 to permit a reasonably accurate dating of Galileo's compositions.

At that time the course of studies at the Collegio was clearly prescribed and a fairly standard syllabus was being taught in each of the subjects. The subjects themselves were arranged in a three-year cycle and followed, in the main, the text of Aristotle. The first year was devoted to logic as set forth in the *Organon* and concluded with a detailed study of the *Posterior Analytics*; the second year focused on natural philosophy, covering the *Physics*, the *De caelo*, and the *Meteorology*; and the third year, after concluding the study of natural philosophy with the *De generatione*, treated the *Metaphysics* and the *De anima* to complete the cycle. Usually a professor would begin with a class in the first year and then take that class through all three years of the cycle. Occasionally, however, a professor would manifest particular competence in logic or natural philosophy, say, and would be assigned to teach that specialty more than once. As can be seen from the *rotulus*, portions of which are reproduced in *Galileo and His Sources*, p. 7, Ioannes Lorinus filled that function in logic and Antonius Menu in natural philosophy. Both left rather complete sets of notes, which apparently were used by their successors to map out their own lectures. Some selectivity and reordering of the materials is detectable from year to year, possibly reflecting the varying pedagogical abilities of the lecturers, and yet a remarkable uniformity characterizes the teaching as a whole. The resulting repetition of titles and subtitles into which the various courses were divided makes it difficult to identify any one professor's notes as Galileo's source, but a careful study of the wording can reveal varying degrees of similarity and other clues that point to notes dating from a particular year as Galileo's likely exemplar.

One professor, though probably not himself Galileo's source, deserves special mention for the thoroughness of his lectures throughout the three-year cycle and for the fact that he meticulously numbered and dated all of his lectures in the margins of his teaching notes. This is Ludovicus Rugerius, a Florentine, who began the cycle in 1590 and concluded it in 1592, delivering a total of 1088 lectures in the process. All of these lectures are preserved in a series of codices now in the Staatsbibliothek Bamberg. They were probably sent to the Jesuit college there at the behest of Christopher Clavius, then a colleague of Rugerius, who was originally from Bamberg and who may have wished to provide his fellow Jesuits there with a model teaching program. The course of Rugerius's teaching is plotted in Figure 1. The points circled on the plot designate the dates, recorded in his notes, on which he ended one tract of a course and began another; many of these correspond to his finishing a commentary on one of Aristotle's works or on a book within a particular work. The cumulative number of lectures is shown along the ordinate, and the total number of lectures on a particular work is given in parentheses under its title. The abscissa, on the other hand, is divided into months, from November of one year to October of the next.

Since MSS 27 and 46 have been the main concern thus far, as can be seen in Figure 1, their contents correspond to only a small portion of the materials covered each year in these lectures at the Collegio Romano. One of the standard divisions for each course was the treatise, and as already noted MS 27 contains only two treatises, the first of which may be incomplete; MS 46, on the other hand, contains three treatises, more or less complete, and a fragment of a fourth, much of which has clearly been lost. The two treatises in MS 27 derive from the portion of the year-long logic dealing with Aristotle's *Posterior Analytics*, whereas the first two treatises of MS 46 correspond to the matter covered in Aristotle's *De*

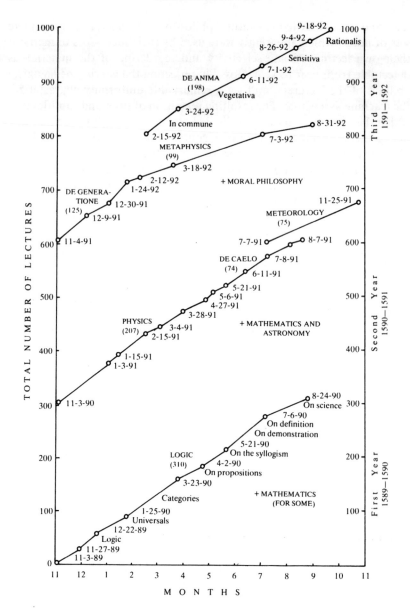

Fig. 1.   Rugerius's Lectures at the Collegio Romano, 1589-1592.

caelo, the second two to the matter covered in his *De generatione*, both from the year(s) devoted to natural philosophy. Since we may presume that professors worked their way through the course at about the same rate as Rugerius, the chronology of his lectures proves useful for dating when these treatises might have been completed in a particular year and thus indicate the earliest time at which they could have been available to Galileo.

Six sets of notes from the course on logic at the Collegio Romano between the years 1584 and 1592 are available to further substantiate the dating of MS 27 as composed not before August of 1588 and most probably in late 1588 or early 1589. Four of these are straightforward records of course work, whose professors and dates of completion are known to be as follows: Ioannes Lorinus (1584), Mutius Vitelleschi (1589), Ludovicus Rugerius (1590), and Robertus Jones (1592). The remaining two are connected with the course completed by Paulus Vallius in 1588, presumed now to be Galileo's exemplar; one is Carbone's plagiarized *Additamenta* of 1597, the other Vallius's reworked logic course as published in 1622. These two versions thus reflect in varying ways Vallius's original commentary on the *Posterior Analytics*. The similarities between them and the contents of MS 27 are so striking that they effectively rule out any other way of accounting for Galileo's organization and precise wording of the materials he there appropriated.

To elaborate somewhat on this analysis, and incidentally to furnish details that prove helpful for dating MS 46, we now give an overview of the contents of all six of the extant logic courses that contain matter similar to that in MS 27. Were we to provide more precise correlations with them as possible exemplars, the proper procedure would be to make comparisons of each of the paragraphs in MS 27 with corresponding materials in the different courses to see precisely how much of the content of each question has counterparts in a possible exemplar. Apart from the content and degree of correlation, the ordering of the questions could also prove important, since the ordering might vary with each reorganization of the subject matter. Such procedures have not proved necessary for determining the source and dating of MS 27, but we mention them here for their later application in the analysis of MS 46.

Lorinus. Table 1 provides a summary of the logic course taught by Ioannes Lorinus and, as already noted, preserved not only in a manuscript dated 1584 but in an almost identical version published at Cologne in

Table 1. Relevant contents of Lorinus's Logic Course on the
Posterior Analytics of Aristotle

Treatise on demonstration

[On foreknowledges and foreknowns]
  On foreknowledges in themselves
    What is foreknowledge
    On the kinds of foreknowledge...
  On the foreknowledges of the third operation [of the intellect]
    On foreknowledge of the subject

Must the existence of the subject always be foreknown	*F3.1*
Can the existence of the subject be proved in a science	
Of the total subject	*F3.2*
Of the partial subject	*F3.4*
Must the quiddity of the subject be foreknown	*F3.6*

    On foreknowledge of the property

Must the existence of the property be foreknown	*F4.1*
Must its real definition be foreknown	

    On foreknowledge of principles

On the foreknowledge of all principles	*F2.1,2.4*
How premises must be foreknown	*F2.2*

[On demonstration]
  On the nature of demonstration

On the definition of demonstration	*D1.1*
What the conclusion must be like	

  [On the conditions of demonstration]

On the first condition: that the premises be true	*D2.1*
On the second condition: that they be...immediate	
What is an immediate proposition	*D2.3*
Must every demonstration contain immediates	*D2.4*
Do axioms enter into demonstrations	*D2.5*
On the third condition: that they be more known	*F4.2;D2.6*
On the fourth condition: that they be prior	*D2.2*
On the fifth condition: that they contain a cause	*D2.2*
On the sixth condition: that they be said of every instance	*D2.7*
On the seventh condition: that they be essential	
On the modes of speaking essentially, in general	*D2.8*
What propositions are in the first...second...	
third... fourth modes of speaking essentially	*D2.9*
What modes...enter into a demonstration	*D2.10*
On the eighth condition: universal, primary, commensurate	*D2.11*
On the ninth condition: that the premises be necessary	*D2.7*
On the tenth condition: that they be proper	

Table 2. Continued.

On the species of demonstration	
How many species are there	*D3.1*
On demonstration of the reasoned fact	
On demonstration of the fact	*D3.2*
On the properties of demonstration	
Is there a demonstrative regress, and what it is	*D3.3*

1620. (A more detailed outline of his course, with the precise Latin titles and the respective foliation and pagination for the various questions in the two versions, is to be found in the Latin Edition, pp. xx–xxiii.) Lorinus was born at Avignon in 1559 and, after entering the Society of Jesus, taught logic at the Collegio during the academic year 1583–1584. Later he went on to a professorship in Scripture there and served as a *censor librorum* within the Jesuit order. He lived until 1634. In Table 1 the questions he treated in his logic lectures that correspond to questions in Galileo's MS 27 are shown in italics, with the number of Galileo's parallel question indicated in the column on the right. This is an abbreviated listing; the full titles of the questions with their subdivisions are in the Latin Edition, noted above. Even a quick perusal of Table 1 and a comparison of it with our Table of Contents will show that a substantial portion of Galileo's material had already been covered by Lorinus.

Another noteworthy feature of the course, though not apparent in the list of questions, is Lorinus's citation of authorities, opinions, and positions. These are practically identical in both versions, indicating that he had added no new sources in the intervening years; Vallius made a similar claim for his two versions, as noted above. Lorinus cites St. Thomas and his followers frequently, as one might expect in view of the long Thomistic tradition at the Collegio Romano. Of Thomists alone there are 178 citations in the exposition of the *Posterior Analytics*, with 69 to Thomas de Vio Caietanus, 26 to Soncinas, 22 to Soto, 18 to Dominicus de Flandria, 11 each to Capreolus and Ferrariensis, and so on. More surprising is Lorinus's even more extensive knowledge of the teachings of the Peripatetics in the Italian universities. Here there are 235 citations in all, with Balduinus having 46 citations, Paulus Venetus 43, Zabarella 31, Niphus 23, and Zimara 19, to mention only the most frequent. Compared to these numbers he is quite sparing in his use of nominalists, with only 33 citations, Jesuits with 8, and Scotists with 7. The Jesuits to whom he refers, predictably, are only two in number: Toletus and Pererius.

Carbone. Carbone's *Additamenta* and related publications give the best indication of the contents of Paulus Vallius's original course, taught four years after Lorinus's, in 1587–1588. The plagiarism itself is quite fortunate, for there are no known manuscripts of Vallius's lectures on logic, and if Carbone had not preserved Vallius's thought the latter's teaching would be unknown to the present day. Table 2 gives an abbreviated listing of the questions treated in the *Additamenta*, attributed, as already explained, to Vallius-Carbone; the full list is in the Latin Edition, pp. xxxii–xxxiii.

Unlike Lorinus's course, which treated foreknowledges as well as demonstration, the *Additamenta* concentrates on foreknowledges alone, containing only one question on demonstration (D1.2), and that incidentally. The reason for this is that Carbone, as Vallius himself was aware, did not plagiarize the treatise on demonstration, whereas he did plagiarize the treatise on foreknowledges and foreknowns. The one question on demonstration happened to be in Vallius's original treatise on instruments of scientific knowing, and since Carbone plagiarized that treatise also, he incidentally preserved the question on demonstration. In Table 2 questions corresponding to those in Galileo's MS 27 are again shown in italics, and our designations for them are indicated on the right. Even a casual perusal of the italicized questions, comparing them with their counterparts in our Table of Contents, should induce one to take seriously Vallius's claim of plagiarism in his *Logica* of 1622. Detailed comparisons of individual questions that establish that claim beyond a reasonable doubt will be found in the Latin Edition, in the Notes and Commentary section that follows the transcription of Galileo's text.

What is most remarkable about the italicized questions in the *Additamenta* is that they contain practically the entire content of Galileo's first treatise, with many passages showing almost word-for-word agreement with Galileo's text. This circumstances makes possible an estimate of precisely how much of Vallius's original notes were appropriated by Galileo. The general idea behind such an estimate is that Galileo's MS 27, which probably dates from 1588–1589, and Carbone's *Additamenta*, published in 1597, both derive from a common source, namely, Vallius's logic course of 1587–1588. If this is so, then a quantitative comparison of the Vallius-Carbone materials in the *Additamenta* with Galileo's materials in MS 27 can be expected to yield a clue to the extent of Galileo's appropriation.

Table 2. Relevant Contents of Vallius-Carbone's *Additamenta*

Treatise on foreknowledges and foreknowns

[On foreknowledges and foreknowns in themselves]
What is foreknowledge, a foreknown, and the ways of foreknowing
How many foreknowledges are there...
Is foreknowledge of a real definition...to be required   *F3.6*
How many foreknowns are there

[On the foreknowledge of principles]
Must nominal definitions of principles be foreknown   *F2.2*
Must principles be foreknown to be true   *F2.1*
How must principles be foreknown, actually or potentially   *F2.3*
In what ways may principles be proved,
   and why are they denied by some
Does a science presuppose its principles in such a way
   that they are never susceptible of proof   *F2.4*

[On foreknowledge of the property]
Must the nominal definition of a property be foreknown
Must the existence of a property be foreknown   *F4.1*

[On foreknowledge of the subject]
On difficulties concerning foreknowledges of the subject
What does Aristotle mean by the word "is," the "is"
   of essence or that of existence?   *F3.1*
Can a science prove the existence of its adequate
   and total subject   *F3.2*
Can the existence of a partial and less principal subject
   be proved in its own science   *F3.4*
Can a science seek the quiddity of its own subject and
   manifest its existence with apodictic proof   *F3.5*

[On foreknowledge of the conclusion]
Are the foreknown major and minor premises known at the
   same time as the conclusion   *F4.2*

[On demonstration]
...
Which instrument is more important, definition or
   demonstration   *D1.2*

Such a comparison is presented in Table 3, entitled "Galileo's Abbreviation of Vallius's Lecture Notes." The following assumptions were made about Carbone in order to simplify the calculations on which the comparison is based: (1) that he preserved the content of Vallius's

Table 3. Galileo's Abbreviation of Valius's Lecture Notes

Identification of question in Galileo's MS 27	Total words in Vallius-Carbone for the comparable question	Words in portions of Vallius-Carbone's question selected for summary by Galileo, with % of (2)	Words actually summarized by Galileo, with % of (3)	Percent of total
(1)	(2)	(3)	(4)	(5)

TREATISE ON FOREKNOWLEDGE:

Foreknowledge of principles:

F2.1	767	486 (63%)	314 (65%)	41%
F2.2	1344	467 (35%)	275 (59%)	20%
F2.3	1344	378 (28%)	218 (58%)	16%
F2.4	1289	712 (55%)	384 (58%)	30%
		(Average: 43%)	(Average: 58%)	(Av. 25%)

Foreknowledge of the subject:

F3.1	2617	1189 (45%)	954 (80%)	36%
F3.2	1244	783 (64%)	646 (83%)	52%
F3.4	1017	722 (71%)	512 (71%)	50%
F3.5	550	500 (91%)	339 (68%)	62%
F3.6	894	566 (63%)	294 (52%)	33%
		(Average: 59%)	(Average: 73%)	(Av. 43%)

Foreknowledge of the property and the conclusion:

F4.1	883	578 (65%)	422 (73%)	48%
F4.2	1683	828 (49%)	755 (91%)	45%
		(Average: 55%)	(Average: 84%)	(Av. 46%)

TREATISE ON DEMONSTRATION:

The nature and importance of demonstration:

D1.2	4950	3533 (71%)	2227 (63%)	45%
		(Average: 71%)	(Average: 63%)	(Av. 45%)
		Overall: 58%	Overall: 68%	Overall: 40%

lecture notes, with some slight additions of his own; (2) that Carbone's additions, in accordance with Moss's findings, were mainly in the area of exemplification or illustration; and (3) that Carbone's editorial preoccupation, as Vallius himself was aware, was one of reordering the materials in his exemplar so as to achieve a better pedagogical result. With regard to Galileo, additional assumptions were made, namely, (4) that he, like Carbone, worked from Vallius's original lectures; (5) that Galileo's aim, as distinct from Carbone's, was one of appropriation for his personal use, which induced him to abbreviate the content of his exemplar rather than develop it; and (6) that Galileo first blocked out areas in Vallius's content, with some rearrangement and renumbering of the questions, before preparing his precis.

These assumptions lie behind the column headings for the data presented in Table 3. The figures reproduced there suggest that Galileo reduced the content of his exemplar by about three fifths, preserving only two fifths of the materials originally available to him. This would not be slavish copying, but would require intelligent summarizing and reorganization, for which abundant evidence is available to anyone who makes a detailed comparison of Galileo's text with the corresponding questions in Vallius-Carbone.

Vitelleschi. Mutius Vitelleschi taught the course in logic at the Collegio Romano during the academic year 1588–1589, at which time he was 25 years of age, having been born in Rome in 1563 and entering the Society in 1583. He continued to teach the cycle of courses in philosophy for the next two years, lecturing on natural philosophy in 1589–1590 and on metaphysics in 1590–1591. Later Vitelleschi became professor of theology, then prefect of studies, and finally was elected General of the Society in 1615. He served in the latter capacity until his death in 1645, during most of the period during which the Jesuits were involved in controversies with Galileo over the Copernican issue.

The notes for Vitelleschi's logic course are preserved in a manuscript in the Vatican Library. An abbreviated list of the questions he treats is given in Table 4; the full listing is in the Latin Edition, pp. xxxvi–xxxviii. Again the questions he has in common with Galileo's MS 27 are shown in italics, with the corresponding designation on the right. As can be seen by comparing Table 4 with the preceding tables, Vitelleschi covered much the same matter as did Lorinus and Vallius, though generally not in as great detail. He apparently based his course more closely on the text of Aristotle than did the others, for he follows the order of text numbers and inserts his questions at the appropriate places, as one might in a combined commentary and questionary. Perhaps for this reason his treatment of foreknowledges is very brief, with only the equivalent of one of Galileo's

Table 4. Relevant Contents of Vitelleschi's Logic Course

ON THE *POSTERIOR [ANALYTICS]* OF ARISTOTLE
On the object of these books

[On foreknowledges]
What is foreknowledge, and how many kinds are there
On what is usually said about foreknowledge
On the foreknowledge that something exists

[On science]
  ...

[On demonstration]

On the definition of demonstration given by Aristotle	*D1.1*
Must the premises of a demonstration be true, and how	*D2.1*
What does Aristotle mean by the expression "first and immediate proposition"	*D2.3*
What is an immediate proposition	*D2.3*
What are the kinds of immediate proposition...	
Must a demonstration be composed of immediates, and how	*D2.4*
On Aristotle's division: more known to nature and to us	
Does demonstration require more known to nature or to us	*F4.2*
Must the premises of a demonstration be more known than the conclusion, and how...	*D2.6*
Must a demonstration be composed of propositions that are said of every instance	*D2.7*
Does Aristotle give a correct definition of the first mode of speaking essentially	
What propositions are contained...in this first mode of speaking essentially	*D2.8*
What is the second mode of speaking essentially	*D2.8*
What propositions are contained in the second mode...	*D2.8*
On the third mode of speaking essentially	
On the fourth mode of speaking essentially	
The four modes being explained, what ones will enter into a demonstration	*D2.10*
Is every essential proposition necessary... [and is] every necessary proposition essential	
What is a first and commensurate predicate	
What are first and universal predicates	*D2.11*

On the division of philosophy and of all the sciences
  ...

---

questions, F4.2, being included among those on demonstration. On the
other hand his treatise on demonstration is quite full and contains a
goodly number of parallels with corresponding questions in Galileo's
manuscript. We should note that following the treatise on demonstration
Vitelleschi includes in his logic notes a treatise on the division of
philosophy and of all the sciences. This is peculiar in that the course he
offered in natural philosophy in the following years begins with the very
same treatise – a possible indication that he resumed with matter he had
been unable to finish during the year devoted to logic.

Rugerius. Ludovicus Rugerius, the Florentine referred to above,
followed Vitelleschi in the philosophy cycle at the Collegio Romano,
where he taught logic during the academic year 1589–1590, natural
philosophy in 1590–1591, and metaphysics in 1591–1592. All of the
lectures he delivered over this three-year period are conserved in a
manuscript now in the Staatsbibliothek at Bamberg, West Germany. As
already mentioned, the manuscript is unusual in that all the lectures are
numbered in the margins, and the dates also given on which the major
tracts were begun. His logic course is in the tradition of Lorinus, Vallius,
and Vitelleschi, but developed at greater length than Vitelleschi's. Indeed,
of all the surviving lecture notes from this period, Rugerius's come closest
in scope and detail to the *Logica* printed by Vallius in 1622. An
abbreviated outline of the contents of his logic course is given in Table 5;
the full listing will be found in the Latin Edition, pp. xxxix–xliv. Again,
the questions shown in italics correspond to those in Galileo's manuscript,
with our designations for them given on the right. It is noteworthy that
Rugerius has only one disputation, that on demonstration, which he
subdivides into three treatises dealing respectively with foreknowledges,
the conditions of a demonstration, and the species of demonstration,
whereas Galileo, as can be seen from our Table of Contents, has two
treatises, one each on foreknowledges and demonstration, and subdivides
the treatises into three disputations apiece. Another peculiarity of
Rugerius's organization, not shown on Table 5 but seen in the Latin
Edition, is that he divides each of his questions into a number of *quaesita*
or queries, thus facilitating more precise partitionings of his subject
matter.

Jones. Robertus Jones, an English Jesuit who also used the names of
Holland, Draper, and perhaps Northe, entered the Society in Rome in

Table 5.   Relevant Contents of Rugerius's Logic Course

Disputation on demonstration

Treatise on foreknowledges
Whether there are such, in general...
Does each and every teaching come about through preexistent
    knowledge, and how
What are foreknowledges, and what kinds are there

What is foreknowledge of principles and how is it described
For what principles is foreknowledge of existence required   F2.1
Is foreknowledge of definition also required for principles   F2.2
What principles must be foreknown earlier and what later

What is foreknowledge of subject and property, and how is it described
What is meant by the existence of the subject   F3.1
Must the existence of the subject necessarily...be foreknown
Must the nominal or the real definition of the subject
    be foreknown   F3.6
How does the property's definition differ from the subject's
Must the existence of the property also be foreknown   F4.1

Treatise on the conditions of a demonstration
What is a demonstration   F1.1
Whether a demonstration must be made from truths and causes
    Must the premises...be true   F2.2
    Must the premises...be causes   F2.3
Must...a demonstration be made from firsts et immediates
    Must the premises...be immediate [propositions]   F2.4
Must a demonstration be made from propositions that are
    prior and more known   F4.2;D2.6
What is a predicate that is said of every instance   D2.7
What is a mode of speaking essentially, and how many are there
    How are the modes of speaking essentially divided   D2.8
    What is the first mode... the second mode...   D2.9
What is a commensurate universal and what propositions can
    be said to be commensurately universal
Are these all properties of a demonstration, and how   D2.10
Must a demonstration be made from proper [principles]   D2.12
Can there be demonstration of corruptibles and singulars

Treatise on the species of demonstration
How many species of demonstration are there   D3.1
How would one describe demonstration of the reasoned fact
How would one describe demonstration from a remote cause   D3.2
How would one describe demonstration from an effect   D3.2
Whether, why, and how one may have a demonstrative regress   D3.3

1583 and taught logic at the Collegio Romano in 1591–1592, natural philosophy in 1592–1593, and metaphysics in 1593–1594. After this he returned to England, later to head the Jesuit mission there, where he died in 1615. His immediate predecessor in the course on logic was Alexander de Angelis, none of whose notes are known to have survived; before de Angelis, of course, Rugerius, Vitelleschi, and Vallius had all taught logic in the years immediately preceding. Jones's course is preserved in the Biblioteca Casanatense in Rome in Codex 3611, consisting of 384 folios. Its contents are shown in abbreviated form in Table 6, with questions corresponding to Galileo's in italics and their designations on the right; the full listing is given in the Latin Edition, pp. xliv–xlvi. Jones seems to have based his course on that of Rugerius, for its contents are similar to the latter's, as can be seen from a comparison of the two. His notes are much abbreviated, however, for although his coverage is quite extensive, he does not go into the detail we find in Rugerius.

VALLIUS AS THE SOURCE OF MS 27

With this we come to Paulus Vallius, the presumed source of MS 27 on the basis of the information already given; subsequent details will establish that identification beyond reasonable doubt. Vallius was born in Rome in 1561 and entered the Society of Jesus in 1582. He taught both philosophy and theology at the Collegio Romano, where his name was entered in the *rotulus* of professors as Valla, although he was sometimes referred to as de Valle. He himself preferred the latter spelling, as evidenced by the fact that he used its Latin form, Vallius, when naming himself in the *Logica* of 1622. (We use the Vallius spelling throughout this volume, despite the fact that we and others have referred to him as Valla in the earlier literature on MSS 27 and 46.) While attached to the philosophy faculty at the Collegio Vallius taught the course in logic in 1587–1588, that in natural philosophy in 1588–1589, and that in metaphysics in 1585–1586, 1586–1587, and 1589–1590. In the preface to his *Logica* he mentions that he has commented on all the philosophical works of Aristotle and has his commentaries ready for publication. Apparently his logic course was the only one to be put into print. A censorship report relating to his commentary on the *Physics*, dated September 8, 1621 and signed by Lorinus, who was also a censor for the *Logica*, is still preserved in the Roman Archives of the Jesuit Order. This confirms his plan to proceed with the publication of the entire course – a plan unfortunately interrupted by his death in 1622.

Table 6. Relevant Contents of Jones's Logic Course

Disputation on matters that pertain to foreknowledges and foreknowns	
[On foreknowledges in general]	
What is foreknowledge, and how many kinds are there	
What is meant by foreknowledge of a definition	
On foreknowledges and foreknowns in particular	
Must the existence of the subject always be foreknown	*F3.1*
What kind of existence of the subject must be foreknown	*F3.2*
On foreknowledge of the property	*F4.1*
On foreknowledge of principles	*F2.1*
On other matters pertaining to the teaching on foreknowledges	
Whether and how first principles arise from preexistent knowledge	
By what habit are principles known	*F2.3*
On the knowledge of principles in relation to other types of knowledge	*F4.2*
On other matters pertaining to the teaching on foreknowledges	
Disputation on demonstration	
On the essence of demonstration in general	
What is demonstration	*D1.1*
Must demonstration be made from true [propositions]	*D2.1*
Must every demonstration be made from firsts and immediates	*D2.2*
What and how many are the kinds of immediate proposition	*D2.3*
Must every demonstration be made from propositions that are more known, prior to, and causes of the conclusion	*D2.2;2.5*
Whether and how every demonstration is made from necessaries	*D2.7*
What propositions and how many kinds of them are contained under the first mode of speaking essentially	*D2.8*
What propositions and how many kinds of them are contained under the second mode of speaking essentially	*D2.8*
On the species of demonstration and on the demonstrative regress	
What and how many are the species of demonstration	*D3.1*
Is demonstration of the fact true demonstration	*D3.2*
Is there a demonstrative regress and circular [reasoning]	*D3.3*
What kind of demonstrative middle term is required for most perfect demonstration, namely, that of the reasoned fact	

Apart from the records of his teaching at the Collegio Romano, his *Logica*, and these censorship reports, little has been known about Vallius. Recently, however, while working in the Roman Archives, Mario Biagioli has uncovered a letter from the General of the Jesuits to a Ioannes Sagredo in Venice and dated November 17, 1601.[12] It seems that Sagredo had complained to the General because of the removal of a professor of philosophy, Paolo Valle, from the Veneto, where he presumably had been teaching at the local college, then located in Padua. This is most interesting, for Paolo Valle without doubt is the Paulus Vallius who authored the *Logica* of 1622. What makes it interesting is the fact that around that time two other Jesuits who were adept at mathematical physics were at Padua and were actually in contact with Galileo. One of these was Ioseph Blancanus, who apparently had studied under Clavius; the other was Andreas Eudaemon-Ioannis, who seemingly had studied under Vallius. This increases the probability that Vallius knew Galileo personally, either as the mathematician who obtained his logic notes while Vallius was teaching in Rome or later when Galileo was the mathematics professor at Padua.

The two-volume revision of Vallius's logic course is described in some detail in *Galileo and His Sources*, pp. 20–23. The portions of this work that cover the same ground as Galileo's logical treatises are found in the second volume, mainly in the disputations on foreknowledge and on demonstration. The titles of the relevant questions are many; their entire listing in the Latin Edition covers fourteen pages, pp. xlvi–lx. A list in English translation, highly abbreviated, is shown in Table 7. This abbreviated form makes it difficult to understand Vallius's assertion in the preface to the second volume of his *Logica* that he has so expanded the logic course of 1587–1588 that no one but the author of the two versions would recognize the elements they have in common. Yet, when we compare the contents of the treatise on foreknowledges and foreknowns in the *Additamenta* with Vallius's first disputation in his exposition of Book 1 of the *Posterior Analytics* in the *Logica* (VL2: 136–166), we find that essentially the same matter has been covered, the same *sententiae* taken into account, and the same conclusions reached. This must also be said of the question that compares definition with demonstration, conserved in the *Additamenta* in the treatise on instruments of scientific knowing (fols. 23va–31va), and originally contained in Vallius's treatise on demonstration (cf. Galileo's D1.2). In the expanded version Vallius takes up this precise problem and defends the same conclusions, only he does so, not in the disputation on foreknowledges or in that on demonstration, but in the disputation on definition (VL2: 377–409). And finally, when we compare Galileo's questions on demonstration that are not conserved in the *Additamenta* with the fuller treatment in the *Logica* (VL2: 188–349), we see that all of these questions are also treated there,

Table 7. Relevant Contents of Vallius's *Logica* of 1622

Disputation on foreknowledges
  On foreknowledges in general
    What is foreknowledge, in general
    Is the foreknowledge of definition that of the nominal
      definition only, or of the real definition also   *F3.6*
    How many kinds of foreknowns are there
  On foreknowledge of principles...
    Must the nominal definitions of terms occuring in principles
      be foreknown prior to demonstration   *F2.2*
    Must principles be foreknown to be true   *F2.1*
    Is it necessary that they be foreknown habitually...   *F2.3*
    How first principles...can be proved   *F2.4*
    Whether and how assent to the premises is included in assent
      to the conclusion   *F4.2*
  On foreknowledge of the property...
    Must the existence of the property be foreknown prior to
      demonstration   *F4.1*
  On foreknowledge of the subject
    Is the existence of the subject always supposed in a science   *F3.1*
    What are the conditions of the subject or object of a science   *F3.2*
    How partial subjects can be proved [to exist] in a science   *F3.4*
    What kind of existence must be foreknown of the subject...
      The "is" of essence must be foreknown of the subject   *F3.5*
Disputation on demonstration
  On the principles of demonstration, in general
  On the nature and conditions of demonstration
    What is demonstration   *D1.1*
    Must demonstration proceed from true [propositions]...   *D2.1*
    What propositions are immediate and first   *D2.3*
    Must true demonstration proceed from immediates   *D2.4*
    [Axioms] do not enter into...each demonstration   *D2.5*
    Must demonstration proceed from causes...   *D2.2*
    Must demonstration proceed from more knowns and priors...   *D2.6*
    What is a proposition said of every instance...   *D2.7*
    On the modes of speaking essentially, in general   *D2.8*
    What propositions are in the first mode... the second mode...   *D2.9*
    Which of these...serve the needs of demonstration   *D2.10*
    Is every necessary proposition essential, and vice versa
    What is a first or commensurate predicate, and its kinds   *D2.11*
    Must demonstration proceed from necessary [propositions]...   *D2.12*
  On the species of demonstration
    How many kinds of demonstration are there   *D3.1*
    Is demonstration of the fact true demonstration   *D3.2*

Table 7. Continued.

Is demonstration of two kinds: demonstration of the fact and demonstration of the reasoned fact	
On circular reasoning and the regress	
What is circular demonstration and can it be admitted	*D3.3*
What is the demonstrative regress and how is it done	*D3.3*
Is a demonstrative regress of this type to be admitted	*D3.3*
Disputation on definition...	
Is definition superior to demonstration	*D1.2*

again with the same discussion of opinions and the same decisions reached.

Correlations and Comparisons. The correlations that can thus be established between Galileo's treatises, the portions of Vallius's notes of 1587–1588 conserved in Carbone's *Additamenta* of 1597 (thus Vallius-Carbone), and the fuller exposition in Vallius's *Logica* of 1622, are displayed in Table 8. Such tabulations not only confirm Vallius's charges in the prefaces of his *Logica*, but they offer substantial grounds for affirming that Galileo's exemplar was a set of Vallius's lecture notes from the course completed at Rome in 1588, and thus similar in all respects to that used by Carbone when preparing his *Additamenta*. Other Jesuit authors such as Lorinus, Vitelleschi, Rugerius, and Jones, treat materials similar to those found in Vallius's two versions of his logic course, but in no case is the agreement as extensive or as close as it is with Vallius's texts. This evidence, taken in conjunction with the data presented in Table 3 *supra*, thus points to Vallius as the exemplar behind MS 27, which was abbreviated and appropriated by Galileo in his own hand, with all the evidence of note-taking pointed out in the commentary on the Latin Edition.

With this conclusion in hand, further light may now be cast on the relationships between the notes appropriated by Galileo and the tradition of logic courses at the Collegio Romano from Lorinus to Rugerius as these have been surveyed in the previous section. Without going into the detail exhibited in Table 8, we may convey a general idea of the agreement by indicating, for each of Galileo's questions, the corresponding questions treated in the other courses, those of Vallius-Carbone and Vallius included. This information is tabulated in Table 9, where an X is used

Table 8. Textual Correlations for the Logical Questions: Galileo, Carbone and Vallius

No. of question	Galileo MS 27	Vallius-Carbone (1588) *Additamenta* (1597)	Vallius *Logica* (1622)
TREATISE ON FOREKNOWLEDGES AND FOREKNOWNS:			
On foreknowledges of principles:			
F2.1	4r2	42ra-42va	2:149
F2.2	4v14	40vb-42ra	2:147
F2.3	5v1	42va-43ra	2:150
F2.4	6r4	43vb-44rb	2:150
On foreknowledges of the subject:			
F3.1	6v23	45vb-48ra	2:159,163-165
F3.2	8r19	48ra-49ra	2:160
F3.4	9r24	49ra-50ra	2:161
F3.5	10r13	50ra-50va	2:164
F3.6	10v13	38rb-39ra	2:164
On foreknowledges of property and conclusion:			
F4.1	11r11	45ra-45vb	2:156
F4.2	11v25	55rb-56vb	2:153
TREATISE ON DEMONSTRATION:			
On the nature and importance of demonstration:			
D1.1	13r17		2:220
D1.2	14r15	28va-31va	2:123,406-409
On the properties of demonstration:			
D2.1	17v17		2:221
D2.2	18v2		2:224
D2.3	19v10		2:229
D2.4	20v6		2:235
D2.5	21r16		2:238-240
D2.6	22r7		2:228,250
D2.7	23v5		2:253
D2.8	24r33		2:255
D2.9	26v1		2:268
D2.10	27r18		2:266
D2.11	27v23		2:273,276
D2.12	28r28		2:281
On the species of demonstration:			
D3.1	29r14		2:299
D3.2	30v20		2:307
D3.3	31r6		2:340,343

to signify when a particular question in Galileo's manuscript duplicates material contained in one or other logic course.

One should note in this table that the X's in the column for Vallius-Carbone are registered in bold face type, to indicate that they show practically word-for-word correspondences with Galileo's questions F2.1 through F4.2, and also D1.2. As already mentioned, Carbone did not plagiarize the treatise on demonstration but only that on foreknowledge and foreknowns, which explains why most of his correlations are with the first treatise. Almost by accident, however, he did appropriate one question from Vallius's second treatise, that on demonstration, namely D1.2, and this is a clue that the exemplar he worked from originally contained both treatises. Galileo, it seems obvious, worked from the same exemplar or from a copy thereof.

The last column on the right, that for Vallius's reworked version, is the only column that lists correspondences for all 28 of Galileo's questions. In some questions there is word-for-word agreement, but this is not generally the case, nor is it to be expected, since Vallius himself acknowledges that he changed the wording to differentiate his definitive work from Carbone's plagiarized version. Yet, since the revision was not published until 1622 whereas the lectures were completed in 1588, the question arises whether the entire contents of the second treatise, and particularly the all-important question dealing with the demonstrative *regressus* (D3.3), was present in Vallius's original lecture notes.

An affirmative answer to this question is indicated from a study of the course offered by Lorinus in 1584. As previously noted, Lorinus stands at the head of the teaching tradition in logic at the Collegio just as Menu, as we shall see later, stands at the head of the teaching tradition on natural philosophy. Lorinus offered the logic course from 1583 to 1586 and then, like Vallius, published it, but not until 1620. His lecture notes dated 1584 are still extant, along with his published volume, and we have already remarked that he did not alter his wording at all in the intervening years. Now, in Table 9, it can be seen that most of Galileo's questions in MS 27 have correspondences in Lorinus's lectures of 1584 (24 of 28). The conclusion seems inescapable that Vallius based his lectures of 1588 on those given by Lorinus in the years preceding. Particularly in the second treatise, that dealing with demonstration, there are such strong correlations between Lorinus's lectures and Vallius's final version (15 of 17) that there must have been commerce between them. Simply on the basis of interchanging the first two columns in Table 9 one should be able

Table 9. Correlations of MS 27 wit Logic Courses at the Collegio Romano

GALILEO MS 27	LORINUS (1584) (1620)	VALLIUS-CARBONE (1588-97)	VITELLESCHUS (1589)	RUGERIUS (1590)	JONES (1592)	VALLIUS (1622)
TREATISE ON FOREKNOWLEDGES AND FOREKNOWNS						
On foreknowledges of principles						
F2.1*	X	[X]*		X	X	X
F2.2	X	[X]		X		X
F2.3		[X]			X	X
F2.4	X	[X]				X
On foreknowledges of the subject						
F3.1	X	[X]		X	X	X
F3.2	X	[X]			X	X
F3.4	X	[X]				X
F3.5		[X]				X
F3.6	X	[X]		X		X
On foreknowledges of property and conclusion						
F4.1	X	[X]		X	X	X
F4.2	X	[X]	X	X	X	X
11	9	11	1	6	6	11
TREATISE ON DEMONSTRATION						
On the nature and importance of demonstration						
D1.1	X		X	X	X	X
D1.2		[X]				X
On the properties of demonstration						
D2.1	X		X		X	X
D2.2	X			X	X	X
D2.3	X		X	X	X	X
D2.4	X		X			X
D2.5	X				X	X
D2.6	X		X	X		X
D2.7	X		X	X	X	X
D2.8	X		X	X	X	X
D2.9	X			X		X
D2.10	X		X	X		X
D2.11	X		X			X
D2.12				X		X
On the species of demonstration						
D3.1	X			X	X	X
D3.2	X			X	X	X
D3.3	X			X	X	X
17	15	1	9	12	10	17

Total number of paragraphs in agreement:

28	24	(12)	(10)	18	16	28

*Bold face type indicates word-for-word similarity; totals in parentheses are for only a portion of the course.

to see that Galileo's notes (and thus Vallius's lectures) represent an appropriation, with only slight development, of the materials already taught by Lorinus.

A few remarks are in order about the remaining columns of Table 9. Vitelleschi covered the least matter in his course, omitting the first treatise almost entirely and covering only about a half of the material in the second. Jones was somewhat more thorough in his treatment of the matter in the lectures of Lorinus and Vallius (16 of 28 questions), but even when he discussed a particular question he did so in much sparser detail than either of his predecessors. Rugerius emerges as the best conserver of the tradition, showing the highest number of questions (18 of 28), and covering each of these in significant detail.

Ordering of Questions. Apart from coverage of individual questions, an alternative way of tracing lines of influence is the ordering of questions within a particular treatise, since professors frequently put their imprint on materials by reordering them in various ways. (Vallius, in fact, in his preface to the 1622 version registered his displeasure with Carbone for reordering the questions he had plagiarized.) Table 10 analyzes the same courses as presented in Table 9, only this time listing the questions not in the order found in MS 27 but in that in which the corresponding questions are found in the other authors. Many interesting observations could be made on the changes from one author to another, but for purposes here it may suffice to note that Rugerius again emerges as the best indicator of the order found in Galileo's exemplar. One can see why Vallius was displeased with Carbone's work – even though he himself changed his ordering somewhat in his 1622 version – by comparing Vallius-Carbone's ordering with that of Galileo. Of all the logic teachers, even though he repeated only 18 of the 28 questions at issue, Rugerius best followed Vallius's ordering in his lectures of 1588. This fact has bearing on our parallel analysis of Galileo's MS 46, which now can be much more extensive than that we gave in our *Galileo's Early Notebooks* of 1977.

Before leaving MS 27 it may prove interesting to address the following question: Supposing that Carbone's plagiarism had never been detected, and thus that the key piece of evidence for our dating of MS 27 were missing, how would one assign a date to Galileo's composition? The answer would have to be framed on the basis of the information contained in Tables 9 and 10, with the respective columns and rows for Vallius-

Table 10. The Order of Questions in Various Logic Courses at the Collegio, 1584-1592

Author	Treatise on foreknowledge and forekowns	Total
LORINUS 1584	F3.1 F3.2 F3.4 F3.6 F4.1 F2.1 F2.4 F2.2	8
VALLIUS 1588 CARBONE 1597	**F3.6 F2.2 F2.1 F2.3 F2.4 F4.1 F3.1 F3.2 F3.4 F3.5 F4.2**	11
GALILEO MS 27	**F2.1 F2.2 F2.3 F2.4 F3.1 F3.2 F3.4 F3.5 F3.6 F4.1 F4.2**	11
VITELLESCHI 1589		0
RUGERIUS 1590	F2.1 F2.2 F3.1 F3.6 F4.1	5
JONES 1592	F3.1 F3.2 F4.1 F2.1 F2.3 F4.2	6
VALLIUS 1622	F3.6 F2.2 F2.1 F2.3 F2.4 F3.5 F4.1 F3.1 F3.2 F3.4 F3.5	11

Author	Treatise on demonstration	Total
LORINUS 1584	D1.1 D2.1 D2.3 D2.4 D2.5 D4.2 D2.6 D2.2 D2.7 D2.8 D2.9 D2.10 D2.11 D3.1 D3.2 D3.3	16
VALLIUS 1588 CARBONE 1597	**D1.2**	1
GALILEO MS 27	D1.1 **D1.2** D2.1 D2.2 D2.3 D2.4 D2.5 D2.6 D2.7 D2.8 D2.9 D2.10 D2.11 D2.12 D3.1 D3.2 D3.3	17
VITELLESCHI 1589	D1.1 D2.1 D2.3 D2.4 F3.5 D2.6 D2.7 D2.8 D2.10 D2.11	10
RUGERIUS 1590	D1.1 D2.1 D2.2 D2.3 F4.2 D2.6 D2.7 D2.8 D2.9 D2.10 D2.12 D3.1 D3.2 D3.3	14
JONES 1592	D1.1 D2.1 D2.2 D2.3 D2.5 D2.7 D2.8 D3.1 D3.2 D3.3	10
VALLIUS 1622	D1.1 D2.1 D2.3 D2.5 D2.2 D2.6 D2.7 D2.8 D2.9 D2.10 D2.10 D2.11 D2.12 D3.1 D3.2 D3.3 D1.2	17

Carbone and Vallius blocked out. *Prima facie*, on the basis of Table 9 one would have to favor Lorinus as the likely source, and date Galileo's writing of MS 27 in 1584 or shortly thereafter. Rugerius would be the next best candidate, and his selection would move the likely date back to 1590. Table 10, on the other hand, might give ground for pause, since it shows that Rugerius best duplicates the order of Galileo's composition, and thus there could be reason, admittedly slight, to favor him. In either event the result would not be conclusive, and the best one might do is locate the composition somewhere between 1584 and 1590.

THE TIME AND PLACE OF COMPOSITION OF MS 27

Since so many indications point to Paulus Vallius as the author of the exemplar on which Galileo's MS 27 is based, we must speculate about the manner in which such an exemplar could have gotten into Galileo's hands, and when and how it happened to do so. Unlike Carbone, Galileo did not study in Rome and had no particular connection with the Jesuits. His early intellectual formation was with the monks at Vallombrosa, a monastery school near Florence, and he pursued further studies at the University of Pisa, whither his father had sent him to prepare for a career in medicine. His basic philosophical studies were thus at that university, where he was a student from 1581 to 1585. Toward the end of that time he became seriously interested in mathematics and dropped out of the university without a degree so as to pursue studies in that discipline. It was this interest in mathematics, and particularly in the methodology of proof, that seems to have led him to Vallius and the Jesuits of the Collegio Romano.

Clavius and Vallius. Since the path by which he apparently did so has been explained in *Galileo and His Sources* (pp. 91–95, 223–225), only its main lines need be sketched here. As Galileo achieved proficiency in mathematics he set to work on a treatise on centers of gravity entitled *Theoremata circa centrum gravitatis solidorum*, a draft of which he completed in 1587. He showed copies of this to various mathematicians, and during a visit to Rome in that year left some of its propositions with Christopher Clavius, the mathematician of the Collegio. Fortunately portions of an exchange of correspondence between Clavius and Galileo on the treatise have been preserved, and these show that Clavius had reservations about the logic of one of the proofs offered in it by Galileo, since he suspected it involved a *petitio principii* [GG10: 24–25, 29–30]. Clavius used this particular expression, commonly employed in scholastic logic for a begging of the question or an assuming of the principle it attempts to prove.

The expression is significant, for it and its inflected forms, such as *petere principium* and *peti principium*, occur thirteen times in Galileo's manuscript. In three of these instances Galileo seems to have had difficulty writing the expression, a possible indication of his discomfort with it. The problem it posed relates to what may legitimately be presupposed in a scientific proof, for it is a problem with the *praecognita* or "foreknowns," treated by Aristotle in the first chapter of the *Posterior Analytics*.

The letters between Galileo and Clavius span the period from January to March of 1588, and there is additional correspondence between Galileo and Guidobaldo del Monte on the subject of the *petitio* that extends further, almost to the end of July of that year [GG10: 34–36]. Clavius himself did not have time to go into details, as he informed Galileo, but it is likely that he asked the professor who was teaching the *Posterior Analytics* about the matter, and possibly put him in touch with Galileo. The professor, of course, was Paulus Vallius, who was covering this matter just about the time of the Galileo-Clavius-Guidobaldo correspondence. Vallius completed the course in August of 1588, assuming that he lectured at about the same pace as Rugerius, who, as seen in Figure 1, finished the treatise *De scientia* on August 24, 1590. As he explained later in the preface to his *Logica*, Vallius made his set of notes available to his students shortly thereafter. Carbone, as has been argued above, got hold of one set of these lecture notes. It seems highly probable that Galileo, at Clavius's request, gained access to another. This would explain the detailed comparisons that can be made between Galileo's manuscript and the teachings contained in Carbone's *Additamenta* as well as those reworked later in the *Logica* of 1622. The agreement between MS 27 and the portions of it that have counterparts in the *Additamenta* would be understandable, since both derived from a common source, namely, Vallius's lecture notes. And the agreement between the teachings contained in MS 27 and the portions of it not contained in the *Additamenta* but reworked in the later *Logica* would also be understandable, since both ultimately derived from the same author, who had not changed his views in the interim but had merely explained them "more fully and at length."

Time of Appropriation. It is difficult to ascertain precisely when Galileo became acquainted with Vallius's materials. The earliest date at which the materials appropriated in MS 27 could have been available to him – arguing again on the basis of Rugerius's schedule for treating the subject matter of the *Posterior Analytics* – would have been the first week in July, for Rugerius finished the tracts *De demonstratione* and *De definitione* on July 6, 1590 (*see* Figure 1). The treatise *De scientia* then occupied him another six weeks or so, until August 24th, as explained above. Since that particular treatise is missing from Galileo's manuscript, it could be that Galileo acquired only the portions of the course Vallius had completed by early July. While possible, however, this scenario raises the question of how Galileo could have acquired this portion of the notes, since there is no evidence that he was in Rome at that time. It seems more likely, therefore, that he had to wait until Vallius finished writing up his lectures and had the opportunity to get one or more sets of them copied for his students. In that event, a set might have been sent to Galileo by Vallius either directly or through an intermediary such as Carbone, whom we know was then in Rome. This alternative would have delayed the notes becoming available until the Fall of 1588. Allowing Galileo some time to absorb their contents and to select the portions he wished to appropriate for himself, one would be judicious in fixing the Winter of 1588–1589 as the likely period during which he worked on the composition of MS 27. At that time he was at Pisa, which would explain the Pisan watermark Drake noted as being on sheets of that manuscript. This date also fits in well with the dating of Galileo's other Latin manuscripts, as we are about to explain.

In what follows, therefore, the time and place of composition of Galileo's logical treatises will be assumed to be early 1589 and at Pisa. This dating has important consequences for deciding the types of references to be included in the commentary to the English translation, just as it influenced the Notes and Commentary to the Latin Edition. Of all the Jesuit logic courses that have been listed above, only the courses of Lorinus and Vallius (the latter both as preserved in the *Additamenta* and as expanded in the 1622 *Logica*) could have influenced Galileo's actual composition. It is for this reason that they are the only courses whose parallels are mentioned in the commentary on the translation. The logic courses of Vitelleschi, Rugerius, and Jones were not yet available at that time, and therefore could not have been used by Galileo. They do have value, however, for showing the continuity of the tradition in logic at the

Collegio from 1589 to 1592, and on this account have been taken into account in the Notes and Commentary of the Latin Edition.

MS 46: THE PHYSICAL QUESTIONS

MS 46 is similar in many respects to MS 27, although it is much longer, being composed of 110 folios as opposed to the latter's 31. It too is an autograph, and Galileo's Latinity in it is much improved throughout, suggesting a later composition on the basis that he was more practiced in appropriating notes of this type. Folios are patently missing from MS 46, and this presents a problem to be discussed later. The extant materials, however, fall into three fairly distinct parts. Two of these, constituting the first 100 folios, are made up of treatises similar to those in MS 27. The third part consists of jottings or memoranda on motion that are obviously related to the *De motu* materials contained in MS 71, and whose discussion, on that account, is best postponed to our consideration of that manuscript.

The two parts or sets of treatises that take up the first hundred folios of MS 46 pertain to portions of a course in natural philosophy that deal with Aristotle's *De caelo* and *De generatione* respectively. These two works are in essential continuity within the Aristotelian corpus, and both discuss the elements, though from different points of view. Normally they would be treated as a unit in Renaissance instruction, following the *Physics* and preceding the *Meteorology*, but they could be separated at the Collegio; an example of such separation has already been seen in Figure 1. This perhaps can serve to explain why Galileo's treatises relating to the *De caelo* were written on paper different from those relating to the *De generatione*, a fact already noted by scholars, and possibly were written at different times.

The Questions and Possible Sources. A detailed listing of the questions still extant in the first two parts of MS 46 is shown in Table 11. There are only 25 of these questions, and upper case letters have been assigned to each, following the convention used in their English translation in *Galileo's Early Notebooks*; the number of paragraphs each question contains is shown on the right. Generally the questions are much longer than those in MS 27; one alone, [K], has almost as many paragraphs as the larger of the two treatises in MS 27. Apart from a brief introduction, there are only four treatises in MS 46, the last of which contains two parts. As

Table 11. Contents of Galileo's MS 46

[ON ARISTOTLE'S BOOKS *DE CAELO*] *Folios 4ʳ-54ᵛ*
[Introduction]

What is Aristotle's subject matter in these books...	**[A]**	*21 pars.*
On the order, connection, and title of these books	**[B]**	*9 pars.*
Treatise on the universe		
On the opinions of ancient philosophers on the universe	**[C]**	*9 pars.*
The truth concerning the origin of the universe	**[D]**	*8 pars.*
On the unity and perfection of the universe	**[E]**	*23 pars.*
Could the universe have existed from eternity	**[F]**	*27 pars.*
Treatise on the heavens		
Is there only one heaven	**[G]**	*34 pars.*
On the order of the heavenly orbs	**[H]**	*36 pars.*
Are the heavens one of the simple bodies or are they composed of them	**[I]**	*47 pars.*
Are the heavens incorruptible	**[J]**	*36 pars.*
Are the heavens composed of matter and form	**[K]**	*183 pars.*
Are the heavens animated	**[L]**	*41 pars.*
...*[Folios probably missing]*...		

[ON ARISTOTLE'S BOOKS *DE GENERATIONE*] *Folios 57ʳ-100ᵛ*
[Tractate on alteration]

...*[Folios missing]*...		
[On alteration]	**[M]**	*2 pars.*
On intension and remission	**[N]**	*32 pars.*
On the parts or degrees of quality	**[O]**	*9 pars.*
Tractate on the elements		
On the elements in general	**[P]**	*15 pars.*
On the quiddity and substance of the elements		
On the definitions of an element	**[Q]**	*17 pars.*
On the material, efficient, and final cause of elements	**[R]**	*6 pars.*
What are the forms of the elements	**[S]**	*17 pars.*
Do the forms of the elements undergo intension and remission	**[T]**	*21 pars.*
...*[Material missing between folios 74 and 75]*...		
[On the number and quantity of the elements]	**[U]**	*80 pars.*
On primary qualities		
On the number of primary qualities	**[V]**	*10 pars.*
Are all four qualities positive, or are some of them privative	**[W]**	*17 pars.*
Are all four qualities active	**[X]**	*21 pars.*
How are primary qualities involved in activity and resistance	**[Y]**	*25 pars.*
...*[Folios missing]*...		

[MEMORANDA ON MOTION] *Folios 102ʳ-110ʳ*

already noted the first two treatises, on the universe and on the heavens respectively, treat matters from *De caelo*, and the remaining two, on alteration and on the elements respectively, treat matters from *De generatione*. Folios are definitely missing from the beginning of the treatise on alteration and from the end of the treatise on the elements, and yet more folios are probably missing from the end of the treatise on the heavens. Some idea of the missing matter can be gained from the following survey of the sources on which Galileo's questions in this manuscript are likely based.

Three complete courses from the Collegio are available for purposes of comparison, and three other partial treatments supplement these. The complete courses are those of Antonius Menu, who taught *De caelo* and *De generatione* in 1578; of Mutius Vitelleschi, who taught the same in 1590; and of Ludovicus Rugerius, who did likewise in 1591. The partial treatments include those of Christopher Clavius, from whose *Sphaera* of 1581 or 1585 Galileo's questions [G] and [H] seem to have been appropriated almost word-for-word, and of Paulus Vallius, whose expositions of *De caelo* and *De generatione* have apparently been lost, but who appended a *Tractatus de elementis* to a course he taught on the *Meteorology* some time between 1586 and 1589; this treatise shows strong correlations with Galileo's *Tractatus de elementis*, i.e., with questions [P] through [Y]. (The remaining document is the manuscript of Ludovico Carbone recently uncovered by J. D. Moss, which offers mainly confirmatory evidence and need not be considered here.)

Menu. A listing of the treatises and questions in Menu's lectures on *De caelo* and *De generatione* is shown in Table 12. Those with counterparts in MS 46 are indicated there and in subsequent tables in italics. As already observed, Menu's role in developing the course on natural philosophy at the Collegio was similar to Lorinus's in developing the course on logic, and he is important on that account. His notes have been examined carefully for signs of agreement with Galileo's composition, and the results shown on the right of the table. Of the 183 paragraphs in Galileo's question [K], for example, materials agreeing with 106 of these can be found in Menu's lectures. For some questions, e.g., [C] through [F], shown in bold-face type, a substantial number of paragraphs contain passages with word-for-word similarity. For still others, e.g., [A], [B], [G], [H], and [V], there are no counterparts or correspondences whatever. Yet the gaps in the listings on the right point to large portions of Menu's

Table 12. Relevant Contents of Menu's Course on *De Caelo* and *De Generatione*

	GALILEO	
ON ARISTOTLE'S BOOKS *DE CAELO*	*Paragraph*	
Tractate on the universe	*Agreement*	
On the origin of the universe...opinions...	**[C]**	*9 of 9*
...the truth about it...	**[D]**	*7 of 8*
Is it demonstrable that the universe was made in time		
Could the universe have existed from eternity	**[E]**	*5 of 27*
On the unity of the universe, or, is there only one	**[F]**	*8 of 8*
On the perfection of the universe	**[E]**	*15 of 15*
Tractate [on the heavens]		
On the nature and essence of the heavens		
Are the heavens an element or composed of them	[I]	*35 of 47*
Are the heavens composed of matter and form	[K]	*106 of 183*
Are the heavens incorruptible by their nature	[J]	*17 of 36*
Are the heavens animated...	[L]	*22 of 41*
What is an assisting form, and in what ways...can an intelligence be called an assisting form		
On the accidents of the heavens		
On their quantity and shape		
On the rarity, density, hardness, transparency, and opaqueness of the heavens		
Is there a difference of positions in the heavens...		
On the motion and the subject of motion of the heavens		
Are the heavens moved by an intelligence or by a proper form		
Is the circular motion of the heavens natural		
Do the stars have light of their own or do they receive it from the sun		
On the action of the heavens		
Do the heavens act on lower bodies		
Whether the heavens act through light, and if so, when		
Do the heavens act through influences		
[ON ARISTOTLE'S BOOKS *DE GENERATIONE*]		
[Tractate on generation]		
[On the causes of generation]		
On the subject of generation		
On the form that is acquired through generation		
On the efficient cause of generation		
On the essence of generation		
[Tractate on alteration]		
[What is alteration]		
In what qualities does alteration terminate	[M]	*1 of 2*
Is alteration a continous motion, and if so, how	[N]	*6 of 32*

Table 12. Continued.

	GALILEO
	Paragraph
How does intension and remission come about	[O] *5 of 9*
[Tractate on action and passion]	
What is action and passion...	
Is all action effected through contact, and if so, how	
Can anything act on itself, and if so, in what way	
Is the reflex action of anything on itself to be admitted	
Can a thing act on something similar to itself, and if so, how	
[Tractate on the elements]	
On the elements in general	
On the term 'element' and other terms	[P] *12 of 15*
What science treats of the elements...	
Proposal of matters to be treated...in the following	
On the existence of the elements	
On the final and efficient causes of the elements	**[R]** *4 of 4*
On the matter of the elements	**[R]** *2 of 2*
What are the substantial forms of the elements	[S] *13 of 17*
Do the forms of the elements undergo intension	
and remission	[T] *10 of 21*
On the definitions of elements	[Q] *12 of 17*
[On the number and distinction of the elements]	
[On the quantity, transmutation, and other accidents of	
the elements]	
Can elements be immediately transmuted into each other	
How there is an easier transition in symbol elements...	
On maxima and minima of the elements	[U] *27 of 74*
On rarity and density...	
On the place of the elements	
On the shape of the elements	
On the ratios of the elements	
On the number and purity of the elements	
On alterative qualities	
On the number of qualities	[U] *6 of 10*
Are the four primary qualities real and positive	[W] *10 of 17*
Why are hotness and coldness said to be active	[X] *6 of 21*
How are qualities related in resistance	[Y] *4 of 25*
On the definitions of humid, hot, etc.	
Can both qualities exist in the highest degree...	
The qualities of the intermediate symbol elements	
On motive qualities	
What is gravity and levity	
Whence do gravity and levity arise	
The motive qualities of the intermediate and simple elements...	
What brings about the motion of the elements	
On the violent motion of heavy and light projectiles	

material that Galileo might have incorporated into notes that have since been lost – the missing folios referred to above. These would include substantial amounts relating to the accidents of the heavens and the actions of the heavens on the sublunary world, entire treatises on generation and on action and passion, and a goodly part of the treatise on the elements, particularly the part relating to their motive qualities, with its lengthy discussion of *gravitas* and *levitas* – extremely important for anyone interested, as was Galileo, in tracts *De motu gravium et levium*.

Vallius. Following in chronological order, the next important possible source for Galileo's MS 46 (apart from Clavius's *Sphaera*) is Vallius's *De elementis*. It is difficult to date this treatise, which is preserved at the end of a codex containing Vallius's undated lectures on the *Meteorology*. Following the order of the Aristotelian corpus, the *Meteorology* should be taught after the *De generatione*, but some adjustments were made in that order at the Collegio because of the large amount of natural philosophy that had to be covered in the *Cursus philosophicus*. In 1591, as can be seen in Figure 1, Rugerius squeezed the *Meteorology* in between the *De caelo* and the *De generatione*, lecturing simultaneously on it and the *De caelo* during July of that year. Earlier, in 1578, Menu followed a similar procedure, teaching the *Meteorology* simultaneously with the first portion of the *De generatione*. A *reportatio* of lectures given by Vallius in 1585 under the title *De mixtis inanimatis imperfectis et perfectis*, the first part of which would correspond to the *Meteorology*, is known to have existed at one time but is no longer available. Again, in a promised revision of his entire course on natural philosophy described in the preface to his *Logica* of 1622, Vallius proposed to treat *De elementis* in conjunction with *De generatione*. Piecing these pieces of information together, since Vallius most likely taught *De generatione* at the beginning of the third year of the philosophy cycle (as did Rugerius in 1591), and did so in 1585, 1586, and 1589, we could assign any one of these dates to the treatise on the elements that has been preserved.

Its dating aside, the contents of the treatise are shown in Table 13, arranged in the same format as Table 12. As can be seen from the "Paragraph Agreement" on the right, all of Galileo's questions show substantial agreement with the first two sections of Vallius's *Tractatus*, many of them containing passages with word-for-word agreement. Apart from Galileo's question [U], which is separated from his previous questions by blank pages anyway, none of Vallius's last four sections has

Table 13. Relevant Contents of Vallius's Lectures on the Elements

	GALILEO
Tractate on the elements	*Paragraph*
On the elements in general	*Agreement*
On the essence of the elements	
Are there elements in the order of nature	[P] *14 of 15*
On the efficient cause of the elements	[R] *6 of 6*
On the matter and form of the elements	[S] *15 of 17*
Do the substantial forms of the elements undergo remission	[T] *14 of 21*
On the definitions of the elements in general	[Q] *13 of 17*
On the number and distinction of the elements	[U] *3 of 6*
On the active qualities of the elements	
Are there only four primary qualities	[V] *6 of 10*
Are all primary qualities active...	[X] *10 of 21*
and positive...	[W] *14 of 17*
How are these qualities related in resistive action	[Y] *9 of 25*
On the definitions of these qualities	
Are symbol qualities of the same species...	
Does each quality exist in an element in the highest degree	
Apart from actual qualities are there...virtual also	
On the transmutation of the elements	
Are the elements transmutable into each other	
Are all elements immediately and mutually transmutable	
Is there an easier transition in symbol elements than in those that are not	
Can a third distinct element be made from two elements that are not symbol elements, and how	
On the quantity of the elements	
Are there maxima and minima in elements	[U] *41 of 74*
Are there ratios in the elements with respect to quantity	
What is rarity and density, and how many kinds are there...	
Is quantity acquired anew in rarefaction...	
On the shape and the purity of elements in their own spheres	
On motive qualities	
What is gravity and levity	
Whence do levity and gravity arise	
Are the motive qualities of the intermediate elements simple and different in species from the extreme elements	
On the place of the elements	
Are the elements moved by themselves or by the generator...	
What moves projectiles	
On the elements in particular	
Is there an elementary fire	
On the properties of fire	
Is air hot of its very nature	

Table 13. Continued.

Tractate on the elements	GALILEO Paragraph
Is water larger and colder than earth	
Is earth the heaviest and driest element	
Do air, water, and earth gravitate in their own spheres	
Tractate on perfect compounds	
....	
[Tractate on imperfect compounds]	

a counterpart in MS 46. These sections, like those in Menu, contain important treatments of the transmutation of the elements, their quantity, their motive qualities, projectile motion, and the answers to such questions as "Does air have weight in air?," much of which occupied Galileo's attention in MS 71. They too, therefore, are suggestive of the type of material appropriated by Galileo on the missing folios of MS 46, once present there but since lost. Moreover, of all the notes relating to *De caelo* and *De generatione*, this treatise by Vallius shows the best agreement with Galileo's MS 46, in a few instances as good as the agreement between Vallius's notes on logic and Galileo's MS 27. While far from apodictic, these are persuasive lines of argument that point to Vallius as the most likely source of both Pisan manuscripts.

Vitelleschi and Rugerius, in that order, followed Vallius in covering the materials of *De caelo* and *De generatione* in 1590 and 1591 respectively. Their value, as will be seen, is similar to that noted in Tables 9 and 10 above, where they basically support Vallius's candidacy by preserving vestiges of his notes, notes that presumably were once available to Galileo. The details of their courses are therefore not given here, though they can be seen elsewhere.[13]

Correlations and Comparisons. With this presentation of likely sources of MS 46, truncated as it may be, we are in a position to sum up our analysis of that manuscript along lines similar to that already provided in Tables 9 and 10 for MS 27. Table 14 is the counterpart of Table 9, giving the overall correlations of the various physics courses offered at the Collegio with Galileo's MS 46 just as Table 9 provided correlations between various logic courses and MS 27. Apart from the correlations for Menu and Vallius based on the data contained in Tables 12 and 13 above, however, Table 14 also includes details of the correlations for the

Table 14. Correlations of MS 46 with Physics Courses at the Collegio Romano

GALILEO MS 46	MENU 1578	CLAVIUS 1581	VALLIUS 1585-1589	VITELLESCHI 1590	RUGERIUS 1591	CARBONE 1594
ON THE BOOKS *DE CAELO*						
Introduction						
[A](21)				13	18	X
[B](9)					5	X
On the universe						
[C](9)*	**9(8)**			5	8	
[D](8)	**8(7)**			3	4	X
[E](23)	**23(11)**			13		X
[F](27)	5			**14(4)**	21	
On the heavens						
[G](34)		**34(34)**				X
[H](36)		**36(36)**				X
[I](47)	**35(2)**			29	18	X
[J](36)	17			25	18	X
[K](183)	106			34	28	X
[L](41)	22			**32(3)**	24	X
474	*225(28)*	*70(70)*		*168(7)*	*144*	
ON THE BOOKS *DE GENERATIONE*						
On alteration						
[M](2)	1			1	1	X
[N](32)	2			25	23	X
[O](9)	5			7	6	
On the elements						
[P](15)	**12(1)**		14	2	**11(4)**	X
[Q](17)	12		**13(6)**	3	12	X
[R](6)	**6(4)**		6			
[S](17)	13		**15(2)**	**12(4)**	4	X
[T](21)	**10(1)**		**14(2)**		11	X
[U](80)	**33(1)**		**44(8)**	42	23	X
On primary qualities						
[V](10)			6	4	7	X
[W](17)	**10(2)**		**14(7)**	15	**15(2)**	X
[X](21)	6		10	**15(2)**	**8(4)**	X
[Y](25)	4		9	**14(1)**	7	
272	*114(9)*		*145(25)*	*140(7)*	*128(10)*	
746	*339(37)*	*70(70)*	*145(25)*	*308(14)*	*272(10)*	

Number of questions in agreement:

25	20	2	10	20	21	20

***Bold** face and number of paragraphs in parentheses indicate word-for-word similarity.

materials found in Clavius, Vitelleschi, Rugerius, and Carbone, not reported in this essay. Table 14 differs in another important respect, namely, while Table 9 shows only generic agreement by means of an X, Table 14 records the fine structure of the agreement in two ways. The first is by recording the total number of paragraphs in the author's questions that correspond to Galileo's, taken from listings such as those provided in Tables 11 through 13, and the second by indicating, in parentheses, how many of these paragraphs contain passages that register word-for-word agreement. This type of breakdown is given for all the courses except Carbone's, for which only generic agreement has been shown.

One striking feature of Table 14 is the role played by Menu in setting up the basic course in 1578, quite analogous to that played by Lorinus in his pioneering logic course of 1584. A number of Menu's questions in the portion of the course dealing with the *De caelo* show strong correlations with Galileo's questions [C] through [L], suggesting either that Menu's notes were directly available to him, possibly in a version revised after 1578 (for Menu did teach the *De caelo* again in 1580–1581), or else that Menu's notes were appropriated by a successor and so passed on to him indirectly. In the latter case the successor could have been Mutius De Angelis, who taught the course in natural philosophy continuously from 1584 to 1587, or Iacobus Caribdus, who taught it in 1587–1588, but neither of whose notes are available. More probably, however, it was Vallius, whose lectures of 1589 on *De caelo* and *De generatione* were then passed on to Vitelleschi and Rugerius, as seems to have been the case with his notes of the previous year on the *Posterior Analytics*.

The latter possibility is suggested by the fairly strong agreement in the next two versions of the course on *De caelo* and *De generatione*, taught by Vitelleschi and Rugerius in 1590 and 1591 respectively. In Vallius's case, as can be seen, the correlations are only for the treatise on the elements added to his lectures on the *Meteorology*, corresponding to the correlations for the treatise shown under Vallius-Carbone in Table 9, but not nearly so good as the latter. It seems, therefore, that the *Tractatus de elementis* for which correlations have been made was not Galileo's exemplar for MS 46, but rather the complete set of notes for the lectures Vallius gave on *De caelo* and *De generatione* in 1589. As we know from his revision of his logical notes, Vallius was a prodigious worker and continually revised his teaching materials. If he saw to it that Galileo received his course on the *Posterior Analytics* of 1588 for the note-taking that shows up in MS 27, it seems reasonable to suppose, as will be explained below, that he would want Galileo to have his most recent course on *De caelo* and *De generatione* for the note-taking that similarly shows up in MS 46.

This line of thought would seem to be confirmed by the data presented in Table 15, analogous to those previously given in Table 10. Note here that there are no counterparts for Galileo's question [A] prior to 1590 or for his question [B] prior to 1591, and that question [V] has no precedent prior to Vallius's *De elementis*. Also, despite minor variations in the ordering of the questions, Vitelleschi and Rugerius best preserve Galileo's ordering in MS 46. The early anticipations in Menu's notes of 1578 need not be a sign of Galileo's copying from them directly, any more than the anticipations in Lorinus's logic notes of 1584 were such a sign of Galileo's direct use of Lorinus for MS 27. An indirect influence of these earlier notes would be sufficient to explain all of the correlations. Similarly, the preservation of more extensive correlations in Vitelleschi and Rugerius need not be a sign of Galileo's having copied from these later authors. Thus, arguing *a pari* from the materials presented in Tables 9 and 10, the materials presented in Tables 13 and 14 point to Vallius as the Collegio professor who was likely behind the composition of MS 46, just as we know him to have been the professor who was behind the composition of MS 27.

THE PROBLEM OF DATING MS 46

Earlier we mentioned that Stillman Drake has agreed with the dating established for the writing of MS 27 by the methods outlined above, and that this dating is confirmed by the watermarks he has identified on sheets of that manuscript. In a more extensive study of watermarks on all of Galileo's pre-Paduan writings, Drake has attempted to extend his results to MSS 46 and 71 also.[14] In this enterprise he uses our dating of MS 27 at about 1589 as a reference point, and with that as a keystone constructs a chronological arch extending back to 1584 and forward to 1591, to which he attaches at various points portions of MSS 46 and 71. The results to which he comes for MS 46 differ somewhat from those implied above, for instead of having the entire manuscript written at Pisa *after* MS 27, he conjectures that its three parts were composed at separated intervals, as follows: (1) the treatises relating to Aristotle's *De caelo* were written at Pisa in 1584; (2) those relating to *De generatione* were written at Florence in 1588; and (3) the memoranda on motion were composed at Florence

Table 15. The Order of Questions in Physics Courses at the Collegio, 1578-1594

Author	On the heavens			On generations			Total
MENU 1578		**CDFE**	IKJL	MNO	ORSTQU	WXY	20
CLAVIUS 1581		**GH**					(2)
VALLIUS 1589				*[On the elements]* PRS**TQU**	VWXYU		(10)
GALILEO MS 46	AB	**CDEF**	GHIJKL	MNO	**PQRSTU**	VWXY	25
VITELLESCHUS 1590	A	CDFE	IKLJ	MNO	UYUPQS	WVXY	(20)
RUGERIUS 1591	AB	CD F	IKLJ	MNO	PQ STU	**WVXY**	21
CARBONE	AB	DE	IKLJGH		PQSU	MN WTVX	20

Questions designated by letters in **bold** face type contain paragraphs with passages that show word-for-word similarity; totals enclosed in parentheses are those for portions of courses or for courses that are incomplete.

and Pisa at various times between 1587 and 1590. According to Drake's dating, therefore, the physical questions of MS 46 were written *before* the logical questions of MS 27, and either before or contemporaneously with the memoranda on motion at the end of the manuscript.

Treatises on the Heavens. The most aberrant of these dates is obviously that assigned to the treatises relating to *De caelo,* for which Drake still finds convincing the argument of Antonio Favaro, who in the National Edition labeled these treatises "youthful writings" or Juvenilia and placed them in 1584, while Galileo was still a student at the University of Pisa [GG1: 12]. Favaro did so on the basis of internal evidence, which he erroneously constructed, in our view, from a biblical chronology recorded by Galileo in the last paragraph of question [D]. The details of that chronology are discussed elsewhere[15]; for purposes here the important point to note is that Galileo there gives the interval between the

destruction of Jerusalem and "the present time" as 1510 years. Since
Jerusalem was destroyed in A.D. 70, the only possible dating a scholar can
derive from Galileo's reference to the present time is 70 plus 1510, or A.D.
1580. Now it is not difficult to discount 1580 as the actual time of
Galileo's writing that particular question. Such an early date is ruled out
by Galileo's youth, since in that year he was only sixteen years of age and
had not yet begun his studies at the University of Pisa. It is possible,
however, that the year 1580 may have been indicated in the exemplar on
which the notes of MS 46 were based. If that were the case, one might
argue either that Galileo gave up on his attempt to recalculate the interval
to the actual time of his writing, and so simply copied the interval given
in his source, or else made an error in calculating and wrote 1510 where he
should have written 1520, which would have yielded 1590, a date
corroborated by other substantial evidence.

It is noteworthy that Drake's argument in favor of 1584 is not based on
watermarks, since all one can deduce from watermark evidence is that the
notes were written at Pisa, which agrees with the result intimated above.
Drake holds that the writing of MS 46 was prompted by Galileo's desire
to obtain a position teaching natural philosophy, since "his father had
warned him not to expect support beyond the academic year 1584–1585,"
and thus he felt impelled to provide for his future.[16] Granted that such
might have been his motivation, it seems unlikely that Galileo would have
had access to the materials found in MS 46 at that early date. During his
student days at Pisa (1581–1585) the *De caelo* was taught at the University
only once, in 1583, when the professors were Hieronimus Borrus,
Franciscus Bonamicus, and Franciscus Verinus. None of the writings of
these men bears close resemblance to Galileo's note-taking. The questions
contained in MS 46 are obviously of Jesuit provenance. How can Drake
explain Galileo's contacts with the Jesuits at that early date? Why,
moreover, should Roman Jesuits have been interested in furthering the
academic future of a 20-year old youth in Pisa, one who was not their
student and who gave no promise of advancing the causes in which they
were interested?

Given the basic fact of provenance from the Roman Jesuits – so amply
demonstrated in the tables analyzed above – the key to the dating of MS
46 must continue to be MS 27, for it alone can provide a satisfactory
explanation of Galileo's contacts with the Collegio Romano and his
motivation for appropriating the extensive series of notes contained in the
two manuscripts. As has been argued above, it was Galileo's interest in

mathematics, not his interest in natural philosophy, that brought about the initial interchange.

Two additional pieces of evidence connect MS 46 with MS 27, and show incidentally that MS 46 could not have been written before MS 27, as Drake has maintained. The first is the terminology employed in MS 46, which presupposes a detailed knowledge of Aristotelian demonstrative logic and the requirements of scientific reasoning; the terminology would be unintelligible, and hence unusable, to one unacquainted with the *Posterior Analytics*. Yet, in MS 46, Galileo treats these difficult matters competently, making no notable errors in his appropriation of the materials contained in his exemplar. In two places, moreover, Galileo presupposes knowledge of conclusions he has already proved in MS 27, one where he affirms in D2.11 that there cannot be a science of individuals, another where he makes use of a position defended in F3.4, namely, that the existence of the partial subject of a science can be demonstrated if it is not known to exist. The second is Galileo's competence in Latin composition, which, as already remarked, is poor in MS 27 and quite passable in MS 46. The exhaustive listings of misspellings and ungrammatical syntax in the Latin Edition amply confirm the conclusion to which Favaro had independently come, namely, that the notes in MS 27, while intelligible, are still the work of a neophyte [GG9: 282]. No such judgment need be made of the Latinity in MS 46 or MS 71. The process here, it must be stressed, is irreversible: one improves with practice in Latin composition, particularly when long time intervals do not separate successive writings, and thus the better the Latinity the later the time of writing.

Once we have established that MS 46 was written after MS 27, the next question is when and for what purpose the composition of this longer manuscript was undertaken. If MS 27 occupied Galileo's time in the early part of 1589, it seems reasonable to expect that he would not have started on MS 46 until later that year. In the summer of 1589 Father Filippo Fantoni, the mathematician who had been teaching the *Sphaera* and the *Theorica planetarum* at the University of Pisa, vacated his position and Galileo was appointed to succeed him. In the fall of that year he therefore began to teach mathematical astronomy, a field in which it would be highly desirable to have a competent knowledge of the physical astronomy contained in Aristotle's *De caelo*. Here Galileo's contact with Vallius probably stood him in good stead, for Vallius himself had passed from logic to the course on natural philosophy at the Collegio, and was actually teaching the *Physics* and the *De caelo* in the academic year 1588–1589. Once Galileo had seen the thoroughness of Vallius's teaching notes for the *Posterior Analytics*, it would be natural for him to turn to Vallius for similar expositions of his teachings on the universe and the heavens. Galileo's motivation, on this accounting, would not be to get a position teaching Aristotelian philosophy, as Drake has conjectured; at the time he had no special competence in that area, whereas he had shown himself to be quite good at mathematics, only requiring improvement in the application of that discipline to astronomy to discharge his new duties properly.

If Galileo did contact Vallius for his teaching notes on the *De caelo* in the summer of 1589, a peculiar situation would have developed that might serve to explain the otherwise intractable 1580 dating associated with the biblical chronology found in MS 46, referred to above. At that time, as can be seen in Figure 1, Vallius would not yet have finished his course on *De caelo*, and so could not have sent his own notes to Galileo; he probably had at hand, however, a very good set of notes deriving from his predecessor, Antonius Menu. Now Menu covered *De caelo* in 1577–1578, at which time his lecture notes show very good agreement for 7 of the 8 paragraphs in Galileo's question [D] – the only paragraph missing being the last paragraph containing the now notorious biblical chronology.[17] But, as already mentioned, Menu also taught *De caelo* one last time, in 1580–1581. It seems quite possible that, when revising his lecture notes in 1580, he decided to fix the date of the origin of the universe and so added the chronology. If he did so, Galileo would have received a set of notes in which the "present time" would have been given as 1580, whereas he himself would have been appropriating them late in 1589 or in 1590, in accordance with the dating proposed above. Again, when improving his lecture notes in 1580, Menu could have added the introductory matter contained in Galileo's questions [A] and [B], thus accounting for presence of these two questions in the lectures of subsequent professors.

Before leaving the *De caelo* portion of MS 46, in light of the foregoing conjecture we may also inquire into the provenance of Galileo's questions [G] and [H], which, though having counterparts in Clavius's *Sphaera* of 1581, do not appear in subsequent Jesuit lectures on the *De caelo*. As can be seen in our commentary on MS 46 in *Galileo's Early Notebooks*, in 1977 we had already questioned whether Clavius's textbook was the direct source of Galileo's two questions, since there are copying errors in them that suggest derivation from a manuscript rather than from a printed source.[18] An examination of the *rotulus* of professors at the Collegio has more recently turned up an interesting possibility.[19] Although none of his lecture notes are extant, Mutius de Angelis taught the *Physics* and the *De caelo* three times in the period between Menu and Vallius, that is, in 1584–1585, 1585–1586, and 1586–1587. In those three academic years, in what surely is more than a coincidence, Clavius did not teach mathematical astronomy at the Collegio, having as substitutes Ricardus Gibbone in 1584–1585 and Franciscus Fuligati from 1585 to 1587. In Clavius's absence, and in light of the growing awareness at the Collegio that mathematical astronomy had important bearing on the matters treated in *De caelo*, it seems reasonable to suppose that De Angelis would abbreviate Clavius's treatment of the number and order of the heavenly spheres in his *Sphaera* and incorporate them into his own teaching notes. If De Angelis did this, they could have become a part of the materials in Vallius's possession, which he then would have passed on to Galileo. On this accounting not only questions [A] and [B] but also questions [G] and [H] became available after Menu's lectures of 1578, which explains why they do not show up in Table 13 before Galileo's appropriating them in MS 46.

The structure of MS 46, as diagrammed in Table 11, points to serious problems about its completeness, that is, whether all the folios that made it up originally are still extant or whether a substantial number have been lost. The first part of the codex, up to and including folio 54, written on paper which Drake identifies as having a Pisan watermark, terminates with Galileo's question [L] on the animation of the heavens. As can be seen from the comparable treatments of *De caelo* in the extant lectures of Menu (Table 12), much more material pertains to the matter of *De caelo* than that appropriated by Galileo in these fifty-odd folios. Of particular importance is the study of the elements from the point of view of their heaviness and lightness and the local motions consequent on these motive forces, which materials, although not treated by Menu, receive detailed attention from both Vitelleschi and Rugerius. In *Galileo's Early Notebooks* we established that Galileo's question [L] was complete and self-contained, and thus that his note-taking on *De caelo* could have terminated at the bottom of the verso side of folio 54.[20] It is probable, however, as maintained by Favaro, that the *De caelo* portion of MS 46, as it has survived, is incomplete, that at one time it contained discussions of the accidents of the heavens and how they act on the sublunary regions, topics that must have been of great interest to Galileo as he set out to teach the astronomy contained in the *Sphaera*. It also might have contained questions on *gravitas* and *levitas*, which would lead directly into the *De motu* treatises to be discussed below. This possibility is more conjectural, since it turns on the precise exemplar available to Galileo – whether it contained only Menu's materials or, alternatively, substantial additions from De Angelis and Vallius that were later incorporated into the lectures of Vitelleschi and Rugerius.

Treatises on Generation. However one resolves that question, there can be no doubt that the portions of MS 46 relating to the *De generatione* are incomplete: at the beginning, in the middle (between the folios presently numbered 74 and 75), and at the end. The material missing at the end is very important, for it is there, based on Menu's lectures, that we would expect to find Galileo's treatments of motive qualities and of natural and projectile motion, which would bear directly on the memoranda on motion in the third portion of MS 46 and on the various treatises on motion in MS 71. It is important to note here that we devoted the entire fourth chapter of *Galileo and His Sources* to likely Jesuit counterparts of the missing materials. The results are so concordant with Galileo's early treatises on motion that one of two conclusions seems inescapable: either Galileo appropriated in writing additional questions that are now lost, or, if not, that he worked over the Collegio exemplar so carefully that it became a mental part of his heritage and so exerted a substantial influence on his later writings.

Like this author and others, Drake locates the *De generatione* portion of MS 46 after the *De caelo* portion, and indeed separates them by a substantial interval of time. His main argument for doing so is that the folios on which the *De generatione* portion is written bear a Florentine watermark, different from the Pisan watermark on the *De caelo* portion. Since Galileo's lectures on Dante's *Inferno*, delivered at the Florentine Academy late in 1588, have the same Florentine watermark, Drake dates the *De generatione* portion in 1588 and holds that it too was written in Florence. His general principle is that manuscripts bearing the same watermark "were composed at one place, from one stock of paper, around one period of time."[21] This principle he further applies to one of the treatises *De motu* in MS 71 and argues that it also was composed in Florence in 1588. But his argument here is not fully convincing. Florence and Pisa are not that distant apart, and Galileo traveled back and forth between them many times; there is little reason to assume that on such trips he could not have carried small packets of paper along with him.

Again, the phrase "one period of time" allows for considerable latitude; if he wrote the Dante lecture in late 1588 from a new stock of paper, he could well be using the same stock of paper through 1589 and even into 1590 and beyond.

Unlike Drake's inference, moreover, there is now no need to posit a long interval between Galileo's writing of the *De caelo* and the *De generatione* portions of MS 46. If there was such an interval, it could have been occasioned by Galileo's having to wait for Vallius to finish his latest lectures on *De generatione*. According to calculations based on Figure 1, Vallius lectured on the *De generatione* in late 1585–early 1586, again in late 1586–early 1587, and for the last time in late 1589–early 1590. Assuming that Galileo had worked on MS 27 in early 1589, then continued with the *De caelo* portion of MS 46 in mid- or late 1589, he would not have gotten to the *De generatione* portion until late 1589 or early 1590, precisely when Vallius's latest teaching notes would have become available. And then, regardless of the watermarks on the paper he was using, he would have been at Pisa, teaching full time at the University.

This, to be sure, is not the only scenario one might excogitate. For example, if Vallius sent Galileo an exemplar of the *De caelo* lectures of one of his predecessors, such as Menu, it would probably have been a codex containing that predecessor's *De generatione* lectures also, since the materials are closely related and not infrequently are found in the same codex. Considering the many lacunae in the information available to us, particularly the unavailability of De Angelis's lectures on *De caelo* and *De generatione*, it is almost impossible to evaluate possibilities such as these. Yet we should not overlook the significance of the data that *are* available and that have been presented in Tables 14 and 15, for these show that the most significant correlations with Galileo's questions in MS 46 occur in the lecture notes of Vallius, Vitelleschi, and Rugerius. These are all fairly late, and thus are discouraging for anyone wishing to situate the completion of MS 46 substantially earlier than 1589 or 1590.

Carbone Again. Having mentioned Tables 14 and 15, we may finally call attention to the last entry in both tables, the name of Carbone, which yet further complicates the possibilities. Reference has earlier been made to the previously uncatalogued manuscript turned up by Moss that summarizes lectures on *De caelo* and *De generatione* attributed to him and dated 1594.[22] The contents of the manuscript are extensive: its lectures on *De caelo* are divided into five treatises containing 38 questions in all, ten of which have counterparts in Galileo's MS 46; those on *De generatione* are divided into seven treatises containing 35 questions, another ten of which have counterparts in Galileo's manuscript. Most of Carbone's questions are similar to those treated in the lectures of Menu, Vallius, Vitelleschi, and Rugerius, and there can be little doubt that they derive from these or similar sources. With regard to the questions treated in common by Galileo and Carbone, there are few word-for-word agreements; generally Carbone presents conclusions that are consonant with Galileo's but leaves out the detailed arguments on which they are based. For purposes here, this is not important. What is important is the type of editorial and diffusion activity in which Carbone, by his own admission, was constantly engaged.

To illustrate the point, a codex in the archives of the Collegio Romano contains lecture notes on the eighth book of Aristotle's *Physics* and the first two books of *De caelo* that resemble Galileo's questions in important ways.[23] Several paragraphs in Galileo's question [K], for example, show almost word-for-word agreement with materials in the codex. Surprisingly, the composer of the lectures was Placidus Carosus, a monk at a monastery in Perugia, and the lectures were given in 1589–1590. How could a monk in Perugia have written questions on *De caelo* that were similar to Galileo's and ended up in the Collegio's archives? Now we have a plausible answer: since Carbone was himself for many years a professor in Perugia, it seems highly likely that he was the intermediary through whom Carosus gained access to the Jesuit materials on which his notes were patently based.

This instance prompts an interesting speculation relating to Vallius, Carbone, and MS 46 along lines similar to those of MS 27. Since Carbone was a frequent visitor to the Collegio Romano and was intent on diffusing Jesuit teachings, it is quite possible that he met Galileo during Galileo's visit to Rome late in 1587. If so, and particularly if Clavius had put Galileo in touch with Vallius before he left Rome, it could well be that Vallius-Carbone stand in a relationship to MS 46 analogous to their relationship to MS 27. Carbone then would have functioned as an intermediary between Vallius and Galileo for the materials of MS 46, just as he was for those of MS 27. The dating of Carosus's lecture notes in 1589–1590 lends support to this speculation, for it was precisely around this time that Galileo was interested in the very materials that had attracted the attention of both Carbone and Carosus.

MS 71: THE *DE MOTU* TREATISES

The possibilities, then, for the transmission of the materials in MS 46 are many, and they are no less so for MS 71. The latter, as already noted, is even more complex because of its containing Galileo's original compositions, and thus the type of argument based on word-for-word agreement that has been used heretofore is no longer available for dating purposes. There are clear connections, however, between MS 71 and MS 46, just as there are between MS 46 and MS 27, and when these are studied closely they also lead to plausible conclusions that are somewhat different from Drake's conclusions based on the study of watermarks. The main result to which they come is that MS 71 was written after MSS 27 and 46, in that order, and for the most part at Pisa, before Galileo left there in 1592 to begin his career at the University of Padua.

Memoranda on Motion (MS 46). The key connector, already noticed by I. E. Drabkin and Raymond Fredette, are the memoranda on motion that occupy the last nine folios of MS 46.[24] These are jottings or excerpts from different sources that show up in various ways in the treatises on motion in MS 71. As argued above and in *Galileo and His Sources*, they are closely connected with the folios that are missing from MS 46 and should be seen as a complement to the Jesuit questions on *gravitas* and *levitas*, on the natural motion of heavy and light bodies, and on projectile motion, with which most of the MS 71 treatises are concerned. Drake does not discuss the watermark on these folios, which is different from the watermarks on the *De caelo* and the *De generatione* portions of MS 46, but Crombie identifies it as "a ladder in a shield."[25] This turns out to be irrelevant for dating purposes, as Drake rightly observes, since the folios are pages from an old notebook apparently used by Galileo's younger brother.[26] Yet the memoranda were obviously written in chronological sequence, and thus by comparing them with their counterparts in the various documents making up MS 71 one can gain a fair idea of the order in which the documents themselves were composed.

If the chronology presented above is correct, the memoranda were begun in late 1589 at the earliest, and more reasonably in 1590. At that time Galileo would again have been teaching mathematics at the University, covering the fifth book of Euclid's *Elements* and probably the *Theoricae planetarum*, an astronomical treatise that usually followed the *Sphaera* he had taught in the previous academic year. It is noteworthy that

Jacopo Mazzoni, Galileo's friend and collaborator,[27] was assigned to teach Aristotle's *De caelo* during the same academic year, so they both would have been concerned with similar subject matters. We might wonder, of course, why Galileo would become interested in *De motu* treatises at this particular time, when his main responsibility was mathematics and mathematical astronomy. But in this connection his predecessor at Pisa, Father Fantoni, had himself written a treatise *De motu* while teaching mathematics there, so there was a precedent for this interest. Charles Schmitt has shown, moreover, that Father Fantoni had based his treatise heavily on the *De motu gravium et levium* of Girolamo Borro, one of Galileo's teachers at Pisa, of whom Galileo is critical in a treatise contained in MS 71.[28] These are all minor considerations, but they may help explain why, apart from the stimulation he may have received from the Jesuit notes, Galileo embarked on the enterprise of MS 71 at this time.

Contents of MS 71. The structure of MS 71 is the following: it begins with a single folio containing a plan for *De motu*; then has a 32-folio dialogue on motion; then an incomplete 18-folio treatise on motion; then a complete 64-folio treatise on motion; and finally two folios containing variants of the first two chapters in the complete treatise. Between the dialogue and the incomplete treatise there is a four-folio insertion, *De motu accelerato*, obviously written later; also, between the complete treatise and the variants of its first two chapters, but bound in upside down and backwards, there is a Latin translation of a Greek work by Isocrates – a residue no doubt of Galileo's classical training before entering the University of Pisa and clearly irrelevant to *De motu*.

Although Favaro was of the opinion that the dialogue on motion was the last piece written, all recent interpreters are agreed that it represents Galileo's first attempt at the subject and so should be dated earliest. The setting of the dialogue is in Pisa, and the main internal evidence available for dating is Galileo's mention in it of his *bilancetta*, which he had invented in 1586, and of a Dionigius Fons, who is portrayed in the dialogue as living but is known to have died on December 5, 1590. The dialogue was also begun before the memoranda on motion, since the first entry in the memoranda is a revision of a portion of the dialogue. Considering that a number of points relating not only to local motion but also to the logic of proof reflect a knowledge of the Collegio materials, it seems plausible to date the dialogue in 1589 and see it as written at Pisa contemporaneously with the materials missing from the *De generatione* portions of MS 46. Drake, lacking watermark evidence that would tie its composition to either Florence or Pisa, locates it in Siena between 1586 and 1587[29]; this dating is also possible, particularly in view of the similarity of its arguments to those in the *Theoremata* on centers of gravity dating from that period.

The remaining treatises on motion are much more problematical, with their dating being a subject of dispute between Drake and Fredette.[30] In our *Galileo and His Sources* we followed Fredette's dating, while acknowledging in a note that some elements of Drake's ordering fit in better with the Jesuit materials being analyzed in that work. Briefly put, the problem is whether the incomplete *De motu* preceded the complete *De motu* or was intended as a partial revision of it; related to that problem is where to locate the two-folio revision of the first two chapters. On the basis of watermarks Drake holds that the incomplete *De motu* was written in Florence in 1588, that the revisions were made in Pisa in 1590, and that the complete *De motu* was written at Pisa (and possibly Florence) in 1590–1591. Fredette, on the other hand, finding more parallels for the later memoranda in the incomplete *De motu*, sees it as being written later than the complete *De motu*, in which more parallels for the earlier memoranda are to be found; if the chronological composition of the treatises follows the sequence of the memoranda, as has generally been held, then Fredette has the better of the argument. The two-folio revisions, on the other hand, represent a consistent attempt on Galileo's part to remove *levitas* completely from the work and to replace *leve* by *minus grave*. For Fredette, therefore, the earlier versions had *gravitas* and *levitas* as two independent principles of natural motion whereas the later versions had *gravitas* only, merely assigning to it various degrees. Drake is forced to hold the opposite position, namely, that Galileo opted first for *gravitas* as a single principle and then returned to the Aristotelian insistence on both *gravitas* and *levitas* as dual principles of natural motion. The Jesuits argued the relative merits of both positions and came down on the side of two principles. Thus, if Drake is correct, there could be a more pronounced Jesuit influence in Galileo's treatise than has previously been recognized. Also, in support of Drake, our examination of a microfilm of MS 71 shows that the folios of what has been referred to above as the complete *De motu* are numbered sequentially by Galileo in his own hand, a fact that has hitherto been unnoticed by scholars. Even though Galileo failed to number the chapter headings of the treatise, therefore, this is an indication that he felt he had in it a fully ordered account.

Others of Drake's speculations, however, do not fit so well with the Jesuit lecture materials. For him, the revisions and the complete *De motu* with its logical structure were inspired by Vallius's notes as appropriated in MS 27; the earlier dialogue and the incomplete *De motu*, on the other hand, were not scholastic or Aristotelian, and thus not of Jesuit origin, but derived instead from Galileo's studies of specific gravity and his interest in Archimedes. Again, in Drake's view the early versions of *De motu* had "a theological or metaphysical opening" that was removed in the later versions. Such speculations, however, introduce a false dichotomy between Aristotelian and Archimedean science; both the earlier and the later versions conform to the methodological canons of the *Posterior Analytics*, as already has been illustrated in *Galileo and His Sources*, pp. 235–248. Again, the Jesuits too were aware of specific gravity and the analyses of local motion provided by Giovan Battista Benedetti; their notes, plus Galileo's contacts with Buonamici and Mazzoni, accord equally well with Archimedean influences.[31] Yet again, a more theological cast is found in the *De caelo* lectures of Menu and Carbone than in those of Vitelleschi and Rugerius, and yet all are of Jesuit origin; the "theological opening," therefore, may reflect only a different set of notes from the Collegio or a different portion of the same set.

Finally, returning to the dialogue and its revision (the first item in the memoranda on motion), we note that even there Galileo could not make up his mind on whether *gravitas* and *levitas* were necessary to explain natural motion or whether *gravitas* alone might suffice. Perhaps a certain ambivalence in his thought persisted to the end, and if so, there is no need to decide between the alternative solutions offered by Drake and Fredette. According to our dating, apart from the dialogue most of the material contained in MS 71 was drafted within a year (or a year and a half at best), hardly sufficient time for a significant evolution to have taken place in Galileo's thought. Although he had made a considerable attack on the problem of motion while at Pisa, he himself was probably aware that he had not solved the problem definitively. The solution he sought would not come until his period of extensive experimental activity at Padua, out of which would come the *De motu accelerato* fragment, the cornerstone of the "new science" of motion that would be featured in the *Two New Sciences* of 1638.

For purposes of this introduction it is not necessary to settle these issues apodictically. The mass of evidence summarized above points clearly to the appropriation by Galileo of MSS 27 and 46 from Jesuit sources while at Pisa around 1589–1590, and for his having completed MS 71, also showing Jesuit influences though more indirectly, by 1591–1592, before his move to Padua. These are very helpful indications for working out the details of the logical methodology to which he was committed at that time and which he would attempt to apply to the subject matters then holding his interest.

ADDITIONAL SOURCES

To complement the sources of Galileo's Pisan manuscripts already sketched we complete this survey by providing an overview of additional sources that could have influenced his composition of the manuscripts as well as his later writings. In the case of MSS 27 and 46, since these so clearly derive from the Jesuit teaching tradition it would seem that additional sources could only have influenced the editorial decisions he may have made while appropriating that tradition. For MS 71, on the other hand, other considerations surely apply. In it Galileo openly discusses Borro's work on falling bodies, and many of the ideas expressed therein seem to have been prompted by the work of another of his teachers, Buonamici. In addition there are notable correspondences in MS 71 with the writings of Benedetti, whose work Galileo does not cite there but which was well known to one of his colleagues at Pisa, Jacopo Mazzoni. As already noted, Galileo provides testimony that he was studying with Mazzoni in late 1590, at the very time he was probably composing the manuscript. These are all indications of possible influences deriving from the University of Pisa that would complement those from the Collegio Romano and thus provide a fuller insight into the sources of Galileo's thought.

Buonamici. Correspondences between the contents of MSS 46 and 71 and the teachings of Buonamici have long been recognized and have been analyzed in some detail by Koyré.[32] More helpful for purposes here, however, is Mario Helbing's recently published study of Buonamici's philosophy.[33] This provides the complete background of Galileo's studies at Pisa, a full analysis of the contents of Buonamici's *De motu*, and valuable reflections on his relations with Galileo. Helbing calls attention to the fact that the *De motu* was already completed by 1587, though it was not published until 1591. Its importance derives from the fact that it records the fruits of Buonamici's teaching at the University of Pisa, where he taught natural philosophy first as an extraordinary professor from 1565 to 1570 and then as an ordinary professor from 1570 to 1587. His occasion for putting out the volume was, in Buonamici's own words, "a controversy that had arisen at the university among our students and certain of our colleagues on the motion of the elements."[34] To appreciate the import of this statement one must be aware, as Helbing points out, that professorial lectures were not the only means of transmitting knowledge to students at the time; disputations were an additional component, and many of these seem to have centered on precisely the problems that interested Galileo throughout his life. It could well be, therefore, that Galileo was one of the students to whom Buonamici refers. The colleagues mentioned most certainly include Borro, who published a treatise on the motion of heavy and light bodies in 1575, to which Galileo refers in MS 71, and probably Filippo Fantoni, the mathematics professor whom Galileo succeeded at Pisa in 1589, who left a manuscript on the same subject that shows Borro's influence, as remarked above.

Helbing's thesis is that Buonamici's teaching exerted a substantial influence on the young Galileo, so much so that his own writings reflect a polemic dialogue with his teacher that continued to the end of his life. The subjects and the problems that preoccupied him were all contained in Buonamici's massive treatise, whose technical terminology Galileo took over as his own, even though his investigations led him to markedly different results. Buonamici's project was to write a definitive treatise on motion in general that would explain its many manifestations in the world of nature, one that would use all the resources of philological research on Greek texts and philosophical reasoning to come to knowledge of the truth. Galileo's project, by way of opposition [GG1: 367], was to concentrate on only one motion, essentially that of heavy bodies, and to make a detailed study of that using mathematical techniques to reveal its true nature. Rather than recover the past and be an apologete for Aristotle, as was Buonamici, Galileo was intent on discovery and innovation, on finding a new science that went beyond Aristotle's. But undoubtedly his awareness of the difficulties this would entail owed much to the foundations he learned from Buonamici in the years between 1581 and 1585. In his lectures, Helbing argues, Buonamici probably introduced Galileo to the atomism of Democritus and Lucretius, to Philoponus's critiques of Aristotle's teachings, to Copernicus's innovations in astronomy, to Archimedes and his use of the buoyancy principle to explain upward motion, to Hipparchus's theory of impetus, and to the writings of many others, including those of Clavius and Pererius at the Collegio Romano – references to all of which can be found in *De motu*. And when Galileo returned to Pisa in 1589 to become Buonamici's colleague, this was precisely when Buonamici was seeing his *magnum opus* through the press, and could have been the occasion of further discussions between the two.[35]

Galileo, of course, explicitly rejected many of Buonamici's teachings. Helbing notes that this rejection is best seen in *La bilancetta*, the *De motu antiquiora*, and particularly in the *Discourse on Bodies on or in Water*, where Buonamici's arguments against Archimedes are definitely Galileo's target. Galileo also makes occasional references to his former teacher in terms that are not complimentary, such as his mentioning the size of his volume on motion both in the *Two Chief World Systems* [GG7: 200] and in the *Two New Sciences* [GG8: 190]. Helbing analyzes an extended passage in the former [GG7: 231–232], exhibiting parallels with it in *De motu* and showing how Galileo's arguments against Buonamici continued long after the latter's death.[36]

Despite these negative reactions, however, Helbing also records three areas of substantial agreement between Buonamici and Galileo that turn out to be particularly relevant to our study. The first is the autonomy they both grant to natural science, separating it off not only from metaphysical speculation but also from religion, what Buonamici calls *pietas* or piety. For both, religious affirmations have no bearing on the study of nature, since faith is not concerned with natural matters but with supernatural. A number of ideas in the *De motu* thus find echo in Galileo's statements in his *Letter to Christina* [GG5: 284, 319], where Galileo, like Buonamici, clearly rejects a theory of double truth on the basis that "two truths cannot contradict one another" [GG5: 320].

A second area is the general methodology both employ in their study of motion. Both wish to use a *methodus* to put their science on an axiomatic base, imitating in this the reasoning processes of mathematicians [*De motu*, 3A–B]. Both regard sense experience as the basic foundation of natural science, taking this in a sense broad enough to include experimentation in the rudimentary form it was then assuming at Pisa. And both see causal reasoning and demonstration, with its twofold process of resolution and composition, as the normal road to scientific conclusions. In this connection a number of causal maxims stressed by Buonamici, also found in Jesuit treatises, find counterparts in Galileo's works. These include: God and nature make nothing in vain [GG7: 85,

429]; a cause precedes that of which it is the cause [GG8: 31]; a difference
in the cause produces a difference in the effect [GG7: 471]; for any one
cause there is only one proper effect [GG1: 164, 7: 423, 8: 60]; and a cause
is that which, being present, the effect is there, and being removed, the
effect is taken away [GG4: 55, 112]. All of these passages are identified by
Helbing in his commentary on Buonamici's text.[37]

Probably the most important area of agreement, however, is the status
each accords to mathematics both as a science in its own right and as an
aid in investigating the secrets of nature. Buonamici lists the three
speculative sciences as physics, mathematics, and metaphysics, and insists
that students should begin their study with mathematics, then proceed on
to physics, and ultimately to metaphysics. Again, mathematics for him is
the discipline that can raise one to divine science. It is also a true science
that satisfies the requirements of the *Posterior Analytics*; in justifying this
view Buonamici explicitly rejects the teachings of Pererius, whose
invectives against the mathematicians are well known, and takes a
position similar to that of Clavius and his student Blancanus. And its
demonstrations are not limited to reductions to the impossible, as some
have held, but they include ostensive demonstrations of all three types: of
the fact, of the reasoned fact, and most powerful, making it the most
exact of the human sciences.[38] Despite his rejection of Archimedes,
moreover, Bounamici further accords validity to the middle sciences
(*scientiae mediae*), which he differentiates from pure mathematics. He
lists these as optics, catoptrics, harmonics, astronomy, navigation
(*nautica*), and mechanics, and sees them as valuable adjuncts for the study
of nature.[39] This part of Buonamici's instruction seems to have deeply
influenced Galileo and set him on the course that would bring him
ultimately to Clavius and the Collegio Romano by the route outlined
earlier in this essay.

Borro and Fantoni. Two additional professors at Pisa, Borro the
philosopher and Fantoni the mathematician, seem to have had a less
positive influence on Galileo. The first, Girolamo Borro, came from
Arezzo and taught at the University of Pisa from 1553 to 1559 and again
from 1575 to 1586, though he was absent during the academic year 1582–
1583. His abrasive personality clashed with those of his colleagues,
including Buonamici, and they succeeded in having him dismissed from
the faculty in 1586. He was the author of three small books, one on the
ebb and flow of the tides (1561), known to Galileo and cited by him in the

Two Chief World Systems [GG7: 445–446, 499], another on the motion of heavy and light bodies (1575), also cited by Galileo in MS 71 [GG1: 333, 367], and the third on method (1584), cited by Neal Gilbert in his *Renaissance Concepts of Method.*[40]

Borro seems to have been the type of Peripatetic philosopher against whom Galileo reacted most violently. He was very different from Buonamici in that he took most of his knowledge of Aristotle from medieval authors, especially Averroes in Latin translation; his writings manifest little acquaintance with the Greek text, cite none of the Greek commentators whose works had recently become available, and show him very much opposed to Platonism and to the attempts being made in his day to reconcile Aristotle's ideas with those of his teacher. His anti-Platonism, coupled with his attraction to Averroes, are further revealed in his vehement rejection of mathematics and of the use of mathematical methods in the study of nature. Borro focused instead on the empirical side of Aristotelian philosophy, stressing the importance of observation and experience in uncovering the secrets of nature, and in this respect undoubtedly exerted an influence on Galileo.

This particular influence is seen in MS 71, where Galileo shows his acquaintance with an experiment performed by Borro and described by him in his work on motion [GG1: 333]. The context is an attempt to decide an argument over whether air has weight in its proper place, that is, in air, an affirmative answer to which would go against the teachings of Archimedes and thus would be of interest to Galileo. To solve the problem Borro dropped two objects in air, one having more elemental air in its composition than the other, to see if both would fall with the same speed or if the one with the greater air content would fall faster. As he describes the test, he obtained a piece of lead and a piece of wood of equal weight, as far as he could judge, and projected the two simultaneously from a high window. While he and other parties to the dispute watched, contrary to their expectations the wood reached the ground before the lead. This they tried not once but many times, always with the same result. From this test (and Borro uses the Latin *periculum* and *experimentum* interchangeably when describing it), he concludes that air must have weight in its proper region, since there is more air in the wood than in the lead and the former falls faster through air as a medium.[41] It is in his discussion of this experiment performed by Borro that Galileo mentions his own tests with falling bodies made "from a high tower," presumably the Leaning Tower of Pisa. Charles Schmitt has remarked on the irony of Galileo's most famous experiment being "anticipated some years earlier by one of the most conservative of his own Peripatetic teachers of natural philosophy."[42]

Filippo Fantoni, a Camaldulese monk who taught mathematics at Pisa and whom Galileo succeeded in that post in 1589, is less interesting than either Buonamici or Borro. He published only one book, a treatise relating to the reform of the calendar that appeared at Florence in 1560, and immediately after that began teaching at the University of Pisa, where he continued until 1567. He then got involved in the affairs of his order and did not return to the university until 1582, when he again taught for a seven-year period, ending in 1589. Records of the courses he taught in his second period are preserved in the *rotulus* of professors, and manuscripts of his lectures are still preserved in the Biblioteca Nazionale at Florence. He expounded Ptolemy's *Quadripartitum* and his *Sphaera*, the first and fifth books of Euclid's *Elements*, the *Sphere* of Orontius Finaeus, and Peurbach's *Theoricae planetarum*, generally following the sequence stipulated in the statutes of the university. More important, he left in manuscript form a compendium of geography and astronomy, preserved in both Latin and Italian versions, and two important *quaestiones*, one on the motion of heavy and light bodies and the other on the certitude of the mathematical sciences.[43]

Fantoni's *De motu* is of some significance for the fact that he wrote it not as a philosopher, as did Borro and Buonamici, but while teaching mathematics, and in so doing set a precedent for Galileo to prepare a similar treatise when he took over Fantoni's post. His work on motion is not very profound, and in the main presents the type of Averroist analysis found in Borro's book. The treatise on mathematics seems similarly unoriginal, taking up basically the same positions we have seen defended by Buonamici in his massive text. It is noteworthy, however, that Fantoni's work is also directed explicitly against Pererius. Fantoni argues that mathematics is a true science, that it fills all the requirements of the *Posterior Analytics* for certain knowledge, that it demonstrates through true causes, and that it can even achieve demonstrations that are most powerful – conclusions consonant with those of Clavius and his disciples. With reinforcement on this matter from both Buonamici and Fantoni, it is not surprising that Galileo would hold mathematics in much higher repute than did those, like Pererius, who were intent on denying its truly scientific character. Fantoni seems to have been on particularly good terms with Buonamici, for the latter refers to him favorably in *De motu*, noting that his own knowledge of the theorem that the angle inscribed in a semicircle is a right angle is not nearly as firm as that "of the famous mathematician, Filippo Fantoni."[44]

Mazzoni and Benedetti. Possibly the strongest influence on Galileo from his years in Pisa, however, came not from his professors there but from a colleague he encountered when he began teaching at the University, Jacopo Mazzoni.[45] Mazzoni joined the faculty slightly before Galileo, not as a mathematician but as a professor of philosophy. Already a respected humanist, in 1587 he had published a scholarly work on Dante, was known to Galileo's father, and was soon to become Galileo's close friend. He taught along with Buonamici, covering the main texts of Aristotle's natural philosophy, but at a salary ten times Galileo's. At the time, as explained in earlier sections of this essay, Galileo was probably engaged in appropriating materials from the Collegio Romano notes. In 1588 he had been invited to give two lectures at the Florentine Academy on the dimensions of hell as described in Dante's *Inferno*, the preparation of which might have led him to knowledge of Mazzoni's work on Dante. In any event, as already observed Galileo wrote a letter to his father on November 15, 1590, asking him to send a copy of his *Sphaera* and some volumes of Galen, and assuring him that he was applying himself to study and was learning from Signor Mazzoni, who sends his greeting. The references to the *Sphaera* and to Galen, cited frequently in the physical questions, suggests that by this date Galileo had progressed to his study of the *De caelo* and *De generatione* and was then composing MS 46. Such an activity would cohere well with the courses Galileo would himself be teaching at the university at that time, and the reference to assistance given by Mazzoni makes further sense, since the latter would have covered the *De caelo* and the *De generatione* in the portion of the philosophy cycle he had taught in the previous academic year.

The contact with Mazzoni is significant for underlining another influence on Galileo during his formative years that was different from the Jesuits' and yet compatible with it. Unlike his Pisan colleagues in philosophy Mazzoni was not a monolithic Aristotelian. Although undoubtedly competent in exposing Aristotle's text, he also had Platonic sympathies, and in the summer of 1589 had actually introduced a course in Plato's thought at the university. One of his major interests was comparing Aristotle with Plato, for he had made a concordance of their views in an early work published at Cesena in 1576. His major work on that subject was entitled *In universam Platonis et Aristotelis philosophiam praeludia* (Preludes to a Complete Philosophy of Plato and Aristotle). As its title indicates, it was concerned not so much with conciliating Aristotle's philosophy with Plato's as it was with recognizing and making good use of the particular strengths of each. The treatise did not appear until 1597, though there are indications Mazzoni was working on it over the intervening years. After its publication at Venice, in fact, Galileo wrote to him and remarked how their discussions at the beginning of their friendship was detectable in its composition [GG2: 197]. The work is therefore valuable for the light it sheds on Galileo's studies in 1590, especially since a number of themes in the *Praeludia* are closely related not only to matters discussed in MSS 27 and 46 but also to the focus on mathematics in MS 71.

Two points Mazzoni uses to compare Plato and Aristotle are their attitudes toward the relative merits of physics and mathematics and their respective views on the use of mathematics in physics. With regard to the first, Mazzoni sees Plato as making mathematics superior to physics and Aristotle as improperly inverting that order. Mazzoni himself follows the traditional division of the speculative sciences, noting that the definitions and demonstrations of physics are concerned with matter and motion and involve all four causes: efficient, formal, final, and material. Influenced by Pererius, whom he cites, Mazzoni then argues that mathematics abstracts from matter and motion and thus does not consider material, efficient, or final causes. He disagrees with Pererius, however, in the latter's contention that mathematics is not a true science, since in his view the mathematician demonstrates through a formal cause, though not considering form precisely as related to matter, as does the natural philosopher. From the viewpoint of its object and its higher level of abstraction, moreover, mathematics occupies a higher position than physics, and paves an easy way from the concrete objects of that discipline to the separated objects of metaphysics. As Frederick Purnell interprets Mazzoni, this ranking is based on an underlying sympathy that connects beings at all levels. Thus Mazzoni's view of Plato enables him to put mathematical entities in a central position vis-a-vis the objects of metaphysics on the one hand and physical bodies on the other. As related to the first they stand in a relationship of effect to cause, as related to the second, in the inverse relationship of cause to effect. On this basis, therefore, Mazzoni sees no difficulty employing mathematics to solve problems in physics.[46]

To the question, therefore, whether mathematical proofs are valid in physics Mazzoni portrays Plato as answering in the affirmative and Aristotle in the negative. Like Buonamici he takes a favorable view of the mixed sciences, the *scientiae mediae*, and he is explicit that not only Ptolemy's work pertains to this category but also that of Archimedes. It was Aristotle's shunning the use of mathematical demonstrations in physics, Mazzoni claims, that caused him to err in his philosophizing about nature. As an example he cites Aristotle's teachings on the velocity of falling bodies. In detailing the particular errors and how they can be corrected, he then turns to the work of Giovan Battista Benedetti and particularly to the way he used Archimedean principles to rectify Aristotle's teaching. Yet Mazzoni's resolution of the problem of velocity of fall does not agree completely with Benedetti's, and indeed more resembles that presented by Galileo in MS 71. This gives reason to believe that it was precisely these matters that Galileo and Mazzoni were studying late in 1590, the period during which scholars are commonly agreed that Galileo was working on his *De motu antiquiora*.[47] But the fact that so many of Galileo's own professors and colleagues at Pisa endorsed the use of mathematical reasoning in physics, especially Buonamici, Fantoni, and now Mazzoni, surely gave him incentive to continue the line of research he was beginning to pursue – even in the face of opposition of conservative Aristotelians elsewhere, including Pererius and his followers.

Another point of comparison made by Mazzoni comes from his interest in pedagogy and concerns the relative merits of Plato and Aristotle for removing impediments to learning. In this matter too Mazzoni favors Plato over Aristotle, at least partially because he sees the use of mathematics as enabling one to overcome impediments encountered in the study of nature. As will be seen, Galileo discusses such impediments in MS 27 and the various suppositions that may be used to circumvent them. It is not unlikely that his studies with Mazzoni were also seminal in this regard.

With regard finally to Benedetti's work on falling motion, historians have long suspected a connection between it and the various compositions now contained in MS 71, but have been stymied because Galileo makes no mention of Benedetti in his early writings. In particular, the anti-Aristotelian tone Galileo adopts resonates strongly with the tone of Benedetti's *Diversarum speculationum mathematicarum et physicarum liber* (A Book of Diverse Mathematical and Physical Speculations), printed at Turin in 1585. Since this was available well before 1590 and figures so prominently in Mazzoni's *Praeludia*, it seems reasonable to suppose that Benedetti's text was itself the object of Galileo's study with Mazzoni referred to in the letter of November 15, 1590. As we have pointed out elsewhere,[48] Benedetti identifies his basic disagreement with Aristotle as consisting in his use of mathematical principles and methods in the study of nature, a theme recurring in both Mazzoni and Galileo. Benedetti's work likewise abounds in suppositions and thought experiments, many of which are similar to Galileo's, and he, like Galileo, is particularly intent on discovering the causes of various properties of local motion – what they both call the *verae causae*, the true causes, as opposed to those proposed by Aristotle.

These additional sources thus complement the materials contained in MS 27 and provide a fuller understanding of Galileo's intellectual development in his years at Pisa. His respect for Plato, for example, and his privileging Plato over Aristotle in his later dialogues, are at least partially explicable in terms of his contacts with Mazzoni. But here it should be noted that such an influence deriving from Mazzoni would not necessarily work at cross purposes with the materials Galileo appropriated from the Collegio Romano. Although the Jesuits rejected the central themes behind a Platonic theory of knowledge, they were far from adverse to Platonism as a philosophy. Following the Church Fathers, who saw in Plato's works doctrines compatible with Christian belief in creation and the soul, in not a few matters they actually preferred Platonic teachings to those of the pagan Aristotle. Particularly during the period we are here studying, as Crombie has amply demonstrated, they further saw Platonism as fostering interest in, and the study of, mathematics.[49] Thus Galileo's turning to Clavius and the Jesuits of the Collegio Romano for help, as documented above, should not be seen as a relinquishing of insights developed at Pisa, but rather, in the spirit advocated by Mazzoni, as seeking to complement them by taking the best from both Aristotle and Plato to aid him in his investigations of nature.

THE RELATIONSHIP OF MS 27 TO GALILEO'S SCIENCE

Precisely how Galileo's notes on Aristotle's *Posterior Analytics* further influenced the development of his science is difficult to evaluate. The plausibility of an Aristotelian methodology informing his investigations not only in the earlier years, i.e., up to 1610, when he published the *Sidereus nuncius*, but also in the productive years that followed, has already been argued in *Galileo and His Sources*.[50] The case was made there on the basis of Galileo's continued use of terms that recur in the logical treatises, particularly his insistence on having achieved demonstrations and thus having fulfilled the Aristotelian idea of *scientia*, especially when dealing with questions in mechanics and in the study of local motion. Terminological similarities are notably present in treatises that have a mathematical cast, such as the *Theoremata* on centers of gravity, the *De motu antiquiora* in its later drafts, the questions on mechanics (*Le meccaniche*), the *Trattato della Sfera*, and the *De motu accelerato* fragment that recurs and figures prominently in the *Two New Sciences*. All of these works, except the last, are commonly agreed to have been written before Galileo's discoveries with the telescope, and not a few authorities would also include the *De motu accelerato* fragment among them.[51] Important works written after 1610 similarly make use of the canons explained in the *Posterior Analytics*, the more significant among them being the *Discourse on Bodies on or in Water*, the *Letters on Sunspots*, the *Letter to Christina*, the *Discorso de flusso e reflusso del mare*, the *Two Chief World Systems*, and the *Two New Sciences*.

Probably the most striking feature of Galileo's use of these canons is his detailed enumeration and justification of the various *suppositiones* on which his reasoning is based. To this should be added his predilection for the expression *ex suppositione* when explaining his mode of arguing or demonstrating in a variety of subject matters.[52] The fact that the Latin *suppositio* is a proper translation for the Greek *hupothesis* has led some commentators to propose, anachronistically, that such usage signals his use of the hypothetico-deductive method of modern science. Yet such a method, it is commonly agreed, could never achieve the certainty Galileo claimed for his results. On this account it seems preferable to locate him in a logical context that first sought effect-to-cause relationships in the phenomena he was investigating, and then used the cause he uncovered, or believed he had uncovered, to provide a proper explanation for the phenomena. This is the general procedure of the demonstrative *regressus* invoked by the Paduan Aristotelians and appropriated by Galileo in the last question of his treatise on demonstration [D3.3].

Just how well Galileo understood the technicalities of this procedure may be open to question, for he seems to have been confused at times in his understanding of the resolutive and compositive procedures it involved.[53] There are indications, moreover, that he adapted the *regressus* procedure to accomodate it to his experimental techniques and his extensive use of mathematical reasoning, neither of which was employed to any extent by the Paduan Aristotelians who first perfected the method.[54] For this reason the claim need not be made that Galileo ended where he began. Rather allowance should be made for a certain amount of appropriation and adaptation on his part, which would not negate, his changes notwithstanding, the fact that the basic procedure of the demonstrative *regressus* provided the model on which his arguments or proofs were based.

A detailed justification of this conclusion is presented in *Galileo's Logic of Discovery and Proof*, particularly in the part entitled *Logica utens* or "Logic in Use," where his attempts to establish new sciences of the heavens and of local motion are analyzed in relation to MS 27 and the *Logica docens* or "Logic Teaching" on which it is based. Short of duplicating that analysis here we simply provide a few examples to illustrate how he variously employed the *regressus* and reasoning *ex suppositione* to achieve important scientific results in his studies of the heavens and local motion.

Astronomical Demonstrations. The demonstrative regress, as explained in D3.3 and its accompanying commentary, involves a threefold procedure: first, a *progressus* from effect to cause wherein the cause is materially suspected but not yet recognized formally as the cause of the phenomenon whose explanation is being sought; second, an intermediate stage wherein the suspected cause is examined in all possible relationships to the phenomenon to ascertain whether it actually *is* its cause; and third, a second *progressus* wherein the connection is recognized formally as such and the causal explanation of the phenomenon advanced *propter quid*.[55] This type of reasoning, like syllogistic reasoning, is rarely spelled out explicitly, and yet it is employed regularly in scientific investigations.

An early example, from Galileo's *Trattato della Sfera* but employed earlier by Ptolemy and Aristotle, is the explanation of the phenomenon of the moon's appearances when waxing and waning [GG2: 250–251]. The first *progressus* consists in suspecting "materially" that the crescent and gibbous phases of the moon are caused by the moon's having a spherical shape. With this insight, one then moves to the intermediate stage – an intellectual activity to ascertain whether the moon's shape really is the cause. Here some relevant factors might be that the moon is illuminated externally by the sun, that it is seen from many different angles, and that under these circumstances *only* a spherical shape will cause it to exhibit crescent and gibbous phases. The second *progessus* simply reverses the procedure and demonstrates from the cause, now recognized "formally" as the cause, the various phenomena that appear as its proper effects under the stated conditions.

Galileo's discovery of the mountains on the moon can be seen as a similar application of the same method [GG3.1: 62–75]. Here the first *progressus* leads him to suspect that the shadows on the moon's surface he has observed with the telescope are an appearance caused "materially" by its mountainous terrain. This insight leads directly to the intermediate stage, a period of observational and even experimental activity, to see whether or not this is the proper explanation. (Apparently Galileo constructed a model of the moon, then illuminated it and viewed it from various angles, to see if mountains were the adequate cause.[56]) Once convinced of the adequacy, the second *progressus* simply identifies the cause formally and draws the conclusion that there actually *are* mountains on the moon.

Galileo's discovery of the satellites of Jupiter is another graphic illustration.[57] The discovery of "the four Medicean stars" and their changes of place with respect to Jupiter sets up the first *progressus*, namely, one in which the movements of the newly discovered heavenly bodies are traced "materially" to their being moons of Jupiter. At first this is merely suspected, but the suspicion sets up the second or intermediate stage wherein detailed examination of the seemingly erratic motions leads to the conviction that they result from the bodies' actually revolving around the planet, at different periods corresponding to their distances from its center. This brings on the second *progressus*, wherein these revolutions are recognized "formally" as the proper cause of the changes of position of the new "stars," with the conclusion further implied that they are actually moons of Jupiter.

A final example is Galileo's observation of Venus and his discovery that it revolves around the sun and not the earth [GG7: 349–352]. Again the procedure is much the same. The first *progressus*, undoubtedly suggested to him by Copernicus's system, compares the appearances of Venus as seen through the telescope, say, its apparent magnitude and phases, with a likely cause of those appearances, namely, a possible revolution around the sun. The intermediate stage then checks this out, as it were, with more detailed observations and calculations, to ascertain whether such a revolution is formally the cause of the observed appearances. The final step, the second *progressus*, explicitly identifies this cause and from it demonstrates the properties formally connected with it.

All of these examples, it should be noted, can be recast in syllogistic form to show how a demonstration of the fact (*a posteriori*) is identifiable in the first *progressus* and how a demonstration of the reasoned fact (*a priori* and *propter quid*) is identifiable in the second. The demonstrations are those of a mixed science (*scientia media*) that makes use of physical and mathematical premises to establish its conclusions and so can, with reason, be referred to as a mathematical physics. The physical premises are the more problematic, since they suppose that the appearances seen through the telescope are not optical illusions but represent factual states of affairs. The mathematical premises for the most part are supplied by projective geometry, although they too are based on a supposition, namely, that light rays travel in straight lines and thus that optical phenomena can be analyzed with the aid of geometrical principles. But the remarkable thing is that the conclusions of the arguments just sketched, after initial opposition on the part of some who had difficulty with the optical evidence on which they were based, were accepted in Galileo's day as true demonstrations. And we might add that Galileo's conclusions still command assent in the present day, not as mere theories or hypotheses, but as valid demonstrations on which our knowledge of the solar system continues to be based.

The Study of Motion. More difficult to explain is the route by which one can emply the *regressus* and reasoning *ex suppositione* to achieve apodictic results in the study of local motion. This problem exercised Galileo in his later life, and his writings offer few clues to how he understood the logic involved, although there are no doubts about the claims he made for its validity. It seems that in general he preserved the three stages of the demonstrative *regressus* already explained, except that rather than have the first stage conclude to a cause "materially" suspected, as stated in D3.3.14, he began to think of the cause at the end of this stage simply as "supposed," that is, taken *ex suppositione*. The second stage would then consist of examining all of the relationships between the supposed cause and its effect to see whether the former is both the necessary and the sufficient condition for the latter under appropriate suppositions. Some of these suppositions would be concerned with the removal of impediments, such as friction and resistance to motion, which could be regarded as accidental or adventitious causes that prevent one from arriving at its essential and proper causes. The suppositions themselves would then have to be reasonably justified, either experimentally or by measurement in cases involving physico-mathematical reasoning. If one could conclude this empirical program successfully, then one would have certified the *a posteriori* part of the reasoning and could proceed with deducing, in *a priori* fashion, the results the proper cause entails. This could be done in the fashion of a mathematical treatise, especially when the phenomena investigated admit of physico-mathematical analysis in the manner associated with the *scientiae mediae*.

Two examples may serve to illustrate how this type of *regressus* could work for Galileo, the first in the context of the arguments he offered in the *Two Chief World Systems* and the second in the similar context of the *Two New Sciences*.

The argument for the earth's motion from the tides may be begun, in this view, with the first *progressus* stated in suppositional fashion: if the earth is rotating daily on its axis and revolving annually around the sun, then certain tidal variations will be caused in seas on the earth's surface. The intermediate stage that follows this is crucial, for the alleged cause, the earth's motion, is certainly problematical, going as it does against sense experience and the public and religious sentiment of Galileo's day. In his attempts to certify the reasoning Galileo invoked the so-called barge experiments, his observations of the tides, and a variety of secondary causes (e.g., the depth of the sea beds and the shape of their boundaries) that might account for the deviations he encountered from expected results.[58]

The main question that has occupied Galileo scholars for years is whether or not he himself believed that he had concluded this stage successfully. In the *Letter to Christina* he made claims that would induce one to think he felt he had done so, but there are sufficient qualifications to give pause, and one cannot be sure. A reasonable view would be that by 1615 he himself was not certain that he had solved all the difficulties, but was sufficiently confident that they could be solved that he repeatedly used the expression "necessary demonstration" when referring to his proof in the letter.[59] The *Two Chief World Systems* was written under such circumstances that Galileo could not boldly claim his tidal argument to be demonstrative, although some theologians who examined the book thought he had done so.[60] Others regarded the argument as made *ex suppositione* but as invoking a false cause, just as Galileo had evaluated the principles behind Ptolemaic astronomy.[61] The last view represents the majority opinion to this day.[62] The *vera causa* of tidal variation is now thought to be the moon's motion and lunar attraction, so that even were the earth at rest there would still be tidal variations. But the important point to note is that Galileo's logical methodology was not defective. Had he been able to show that the earth's motion was a necessary and sufficient condition for the tides to occur, he would have been able to conclude the second *progressus* and would have achieved the necessary demonstration he was seeking. Unlike the Ptolemaic arguments, his was not based on the *fallacia consequentis* and was not defective from the viewpoint of his early notes on logic.[63]

The demonstrative force of the arguments developed during the Third and Fourth Days of the *Two New Sciences* to establish a *nuova scienza* of motion are even more difficult to evaluate. Schematically, however, they can be formulated in a single argument that shows how Galileo may have thought them demonstrative in the light of a suppositional understanding of the *regressus*. This argument applies to a ball projected horizontally from the top of a table and then allowed to fall naturally to the floor.[64] The first *progressus* in this case is again expressed suppositionally: on the supposition that the ball undergoes a uniform horizontal motion as a result of the projection and at the same time undergoes a uniform vertical acceleration during the period of its fall, the ball will follow a semiparabolic path to the floor.[65] (Other mathematical properties of the resulting motion, such as satisfying the double-distance rule and the times-squared rule, may also be specified, but these are already implied in the parabolic trajectory.) The intermediate stage here again is the difficult one, and it undoubtedly caused Galileo considerable "agitations of mind."[66] This consists in showing, from a large number of experiments and calculations, that a uniform horizontal motion and a uniform increase in velocity of fall with respect to time is the *only* way to explain these mathematical properties within the accuracy of the observed results. Apart from the problem posed by precision in measurement, the mental examination involved suppositions about accidental *impedimenta*, such as friction and resistance, being either eliminated or reduced to the category of incidental causes that do not alter the "essential character" of the motion.[67] In the long run, Galileo believed that such suppositions were reasonable and that he had concluded this stage successfully, and so could proceed to the second *progressus*. This, in effect, provided him with the principles on which his science of motion could be based, namely, uniform velocity along the horizontal axis and uniform acceleration along the vertical, in the absence of impediments that might perturb the result. Thus he could organize his final treatise along the lines of a Euclidean formal exposition, confident that his empirical foundations could sustain a "new science" of kinematics or dynamics that would be on a par with the science of statics Archimedes had formulated successfully so many centuries earlier.

This analysis is presented here as merely suggestive, since fuller documentation is now provided in *Galileo's Logic of Discovery and Proof*. Yet it should give some idea how Galileo's later scientific investigations, as well as his earlier ones, can be assimilated to the methodological canons spelled out in MS 27. Now that these are being made available in English as well as in Latin, it is expected that Galileo scholars will be able to work through the entire corpus of his works and evaluate them in terms of the canons he himself claims to have used. Perhaps in this way some anachronisms may be eliminated from studies of this "Father of Modern Science," who has been seen methodologically as a Pythagorean, a Platonist, a Humean, a Kantian, even a positivist, but rarely as the Renaissance Aristotelian he most probably was.

NOTES TO THE INTRODUCTION

[1] This work has actually been in progress for two decades, though as yet its results are not widely known. The pioneering efforts of Carugo and Crombie are described in A. C. Crombie, "Sources of Galileo's Early Natural Philosophy," in Bonelli and Shea's *Reason, Experiment, and Mysticism in the Scientific Revolution*, 157–175, 303–305. Related research, reported in fuller detail, is described in Christopher Lewis, *The Merton Tradition and Kinematics in Late Sixteenth- and Early Seventeenth-Century Italy*, Padua: Editrice Antenore, 1980. The translator's early investigations are summarized in *Galileo's Early Notebooks: The Physical Questions*, Notre Dame, Ind.: University of Notre Dame Press, 1977; subsequent studies are detailed in his *Prelude to Galileo* and *Galileo and His Sources*. Edwards' transcription of the autograph of the logical questions (MS 27) on which the Latin Edition is based has, of course, made the present volume possible.

[2] *Indici e cataloghi, Nuova serie V*. La collezione galileiana della Biblioteca Nazionale di Firenze, Vol. 1. Florence: Istituto Poligrafico dello Stato, Libreria dello Stato, 1959, 106–107.

[3] *Indici e cataloghi*, 107.

[4] See Crombie, "Sources of Galileo's Early Natural Philosophy," 304–305, and Drake, "Galileo's Pre-Paduan Writings: Years, Sources, Motivations," *Studies in History and Philosophy of Science*, 17 (1986), 436–437.

[5] For a fuller examination of this question see our "Galileo's Sources: Manuscripts or Printed Works?," *Print and Culture in the Renaissance: Essays on the Advent of Printing*

in Europe, eds. G. B. Tyson and Sylvia Wagonheim, Newark, Del.: University of Delaware Press, 1986, 45–54.

[6] A. Carugo and A. C. Crombie, "The Jesuits and Galileo's Idea of Science and of Nature," presented at a Convegno Internazionale di Studi Galileiani entitled *Novità Celesti e Crisi del Sapere* and held at Pisa, Venice, Padua, and Florence on 18–23 March 1983. This statement and the summary that follows are taken from an abstract published in the *Sommari degli Interventi*, Florence: Banca Toscana, 1983, 7–9. A much expanded version of the paper may be found in the *Annali dell'Istituto e Museo di Storia della Scienza di Firenze* 8.2 (1983), 3–68.

[7] As will be seen in what follows there is good reason to believe that Vallius knew Carbone personally and even may have furnished him with copies of his lecture notes. Thus one may wonder why Vallius refers to him as "some good author" in the preface to his first volume and again as "this good man" in the preface to the second. The expressions could have been irony, but more probably are a sign that Vallius knew Carbone to be dead by the time he published his *Logica* and did not wish to damage his reputation. For a list of Carbone's publications see *Galileo and His Sources*, 12–13 n. 23, and the essays cited in the following two notes.

[8] J. D. Moss, "The Rhetoric Course at the Collegio Romano in the Latter Half of the Sixteenth Century," *Rhetorica* 4.2 (1986), 137–151; "Aristotle's Four Causes: Forgotten *Topoi* of the Italian Renaissance," *Rhetoric Society Quarterly* 17.1 (1987), 71–88; "Rhetorical Invention in the Italian Renaissance," in *Visions of Rhetoric: History, Theory and Criticism*, ed. C. W. Kneupper, Arlington, Tex.: Rhetoric Society of America, 1987, 30--41; and "Dialectic and Rhetoric: Questions and Answers in the Copernican Revolution," *Argumentation*, 5 (1990), 17–37.

[9] J. D. Moss, "Ludovico Carbone's Commentary on Aristotle's *De caelo*," in *Nature and Scientific Method*, Essays in Honor of William A. Wallace, ed. D. O. Dahlstrom, Washington, D.C.: The Catholic University of America Press, 1990, 169–192.

[10] "The Rhetoric Course," 147.

[11] "Aristotle's Four Causes," 78.

[12] Personal communication, Biagioli to Wallace, 8 November 1989.

[13] In our "The Dating and Significance of Galileo's Pisan Manuscripts," *Nature, Experiment, and the Sciences*, eds. T. H. Levere and W. R. Shea, Dordrecht: Kluwer Academic Publishers, 1990, 3–50.

[14] In his "Galileo's Pre-Paduan Writings," noted above (n. 4).

[15] In *Galileo*'s Early Notebooks, 21–24; *Prelude to Galileo*, 217–225; and *Galileo and His Sources*, 89–95.

[16] "Galileo's Pre-Paduan Writings," 437.

[17] *Galileo's Early Notebooks*, 42, 258–259, commentary on par. D8.

[18] *Galileo's Early Notebooks*, 263–269, commentaries on pars. G1, G13, G17, H12, and J24.

[19] See Table 1 in *Galileo and His Sources*, 7.

[20] *Galileo's Early Notebooks*, 276, commentary on par. L41.

[21] "Galileo's Pre-Paduan Writings," 431.

[22] BNF Cod. CL XII, 64 Theatini. Moss summarizes the contents of this manuscript in the essay cited in note 9 above.

[23] Rome, Università Gregoriana, APUG-FC, Cod. 690.

[24] The memoranda have been translated by I. E. Drabkin in *Mechanics in Sixteenth-Century Italy*, eds. S. Drake and I. E. Drabkin, Madison: The University of Wisconsin Press, 1969, 378–387.

[25] "Sources of Galileo's Early Natural Philosophy," 305.

[26] "Galileo's Pre-Paduan Writings," 433–434, table on 440.

[27] That Galileo and Mazzoni were collaborating is clear from a letter written by Galileo to his father on November 15, 1590 (GG10: 44–45); see *Prelude to Galileo*, 227. Possible influences of Mazzoni on Galileo, and particularly the possibility of their joint study of Benedetti's work, are discussed in *Galileo and His Sources*, 225–230.

[28] See Essay 10, "Filippo Fantoni, Galileo Galilei's Predecessor as Mathematics Lecturer at Pisa," in Schmitt's *Studies in Renaissance Philosophy and Science*, London: Variorum Reprints, 1981. The essay appeared originally in *Science and History: Studies in Honor of Edward Rosen*, published in Wroclaw in 1978, 53–62.

[29] "Galileo's Pre-Paduan Writings," 440.

[30] See the bibliography cited in *Galileo and His Sources*, 230–231 nn. 22, 24; also 357–358.

[31] *Galileo and His Sources*, 168–172, 184–202. The Jesuit philosophers may not have known of Benedetti directly, but they do cite Jean Taisnier's plagiarism of his work, as noted on 185. There is no doubt that Clavius was acquainted with him, as detailed in our "Science and Philosophy at the Collegio Romano in the Time of Benedetti," *Atti del Convegno Internazionale di Studio Giovan Battista Benedetti e il Suo Tempo*, Venice: Istituto Veneto di Scienze, Lettere ed Arti, 1987, 113–126.

[32] *Galileo Studies*, tr. John Mepham, Atlantic Highlands, N.J.: Humanities Press, 1978.

[33] *La Filosophia di Francesco Buonamici, professore di Galileo a Pisa*, Pisa: Nistri-Lischi Editori, 1989.

[34] Helbing, *Buonamici*, 54.

[35] *Buonamici*, 352–361.

[36] *Buonamici*, 364–370.

[37] *Buonamici*, 77–85.

[38] *Buonamici*, 86–93.

[39] *Buonamici*, 94–97.

[40] New York and London: Columbia University Press, 1960, 176–178.

[41] The Latin text of this passage is given in our *Causality and Scientific Explanation*, 2 vols., Ann Arbor: University of Michigan Press, 1972–1974, 1: 238 n. 110.

[42] In his entry on Borro in the *Dictionary of Scientific Biography*, ed. C. C. Gillispie, 16 vols., New York: Charles Scribner's Sons, 1970–1980, 15: 45.

[43] Fuller details on Fantoni will be found in Schmitt's article cited in n. 28 above.

[44] Helbing, *Buonamici*, 101.

[45] Information on Mazzoni supplemental to that given in n. 27 above will be found in Frederick Purnell, "Jacopo Mazzoni and Galileo," *Physis* 3 (1972), 273–294.

[46] Purnell, "Jacopo Mazzoni and Galileo," 282–284.

[47] "Jacopo Mazzoni and Galileo," 284–294.

[48] In the essay cited in n. 31 above.

[49] See his "Mathematics and Platonism in the Sixteenth-Century Italian Universities and in Jesuit Educational Policy," *Prismata: Naturwissenschaftsgeschichtliche Studien*, eds. Y. Maeyama and W. G. Salzer, Wiesbaden: Franz Steiner, 1977, 63–94. More recently Corrado Dollo has emphasized the same theme, based on his transcriptions of the lecture notes of

Vallius and Vitelleschi, as he informed the author by mail in December of 1989; see his "L'Uso di Platone in Galileo," *Siculorum Gymnasium* 42 (1989), 115–157, for details.

[50] Part III. Galileo's Science in Transition, 217–349.

[51] Antonio Favaro, for example, regarded the fragment as written in 1604, whereas Emil Wohlwill and Alexandre Koyré both thought it written in 1609. For more details, see *Galileo and His Sources*, 273, n. 105.

[52] *Ad indices*: *Galileo and His Sources*, under "*ex suppositione*" and "suppositions, *suppositiones, supposizioni*"; *Prelude to Galileo*, under "*ex suppositione*" and "supposition (*suppositio*)."

[53] The confusion undoubtedly resulted from the different ways in which resolution and composition were applied in mathematics and in the natural sciences; this matter is touched on in various places in *Galileo and His Sources*, e.g., 119–122, 146–147, 213, 285, 302, and 308. A detailed analysis is provided in Sec. 2.7 of *Galileo's Logic of Discovery and Proof*.

[54] Proposals regarding Galileo's methodological innovations with the *regressus* are offered in *Galileo and His Sources*, 338–347, *Prelude to Galileo*, 150–156, and in *Galileo's Logic of Discovery and Proof*, Chaps. 5 and 6.

[55] On the term *progressus* and its value in tracing the sources from which Galileo's teaching derives, see Lat. Ed., 109–110 and 112–113.

[56] He adverts to this in his *Two Chief World Systems*, GG7: 111–112.

[57] In the *Sidereus nuncius*, GG3.1: 79–96.

[58] In the *Two Chief World Systems*, GG7: 442–489, esp. 456, where he makes reference to the *machina* he had used to investigate the phenomena; he had proposed similar arguments in his preliminary draft, *Discorso del flusso e reflusso del mare*, GG5: 377–395, translated in Finocchiaro, *The Galileo Affair*, 119–133.

[59] This is the conclusion reached by J. D. Moss after a careful analysis of all of Galileo's references to demonstration in the context of the Copernican debates. See her "The Rhetoric of Proof in Galileo's Writings on the Copernican System," in *Reinterpreting Galileo*, ed. W. A. Wallace, Washington, D.C.: The Catholic University of America Press, 1986, pp. 179–204; also her earlier essay, "Galileo's *Letter to Christina*: Some Rhetorical Considerations," *Renaissance Quarterly* 36 (1983), 547–576.

[60] This was the judgment, for example, of Melchior Inchofer; *see* GG19: 349–356.

[61] Simplicio states this position explicitly in the *Two Chief World Systems*, GG7: 462.

[62] A notable exception in Stillman Drake, who still argues for the cogency of the argument from the tides in his *Telescopes, Tides and Tactics*, Chicago: The University of Chicago Press, 1983.

[63] The *regressus*, as Galileo explained it and as it was understood by the Jesuits, required that cause and effect be convertible; see D3.3.14.

[64] The literature in which "table-top experiments" of this type are discussed is indicated in *Galileo and His Sources*, 264–268. To this should be added D. K. Hill's "A Note on a Galilean Worksheet," *Isis* 70 (1979), 269–271; his "Galileo's Work on 116v: A New Analysis," *Isis* 77 (1986), 283–291; and his "Dissecting Trajectories: Galileo's Early Experiments on Projectile Motion and the Law of Fall," *Isis* 79 (1988), 646–668.

[65] The argument is developed by Galileo in his *Two New Sciences* of 1638 (GG8: 273–276), but the experimental work on which the thesis is based seems to have been done many years earlier, around 1608–1609; see the literature mentioned in the previous note, plus R. H. Naylor's "Galileo's Method of Analysis and Synthesis," *Isis* 81 (1990), 695–707.

[66] When outlining his justification for adopting the principle of uniform acceleration on the Third Day of the *Two New Sciences*, Galileo uses this expression (*post diuturnas mentis agitationes*, GG8: 197); earlier he had used the same expression in a draft preserved in the *De motu accelerato* fragment, GG2: 261.

[67] Galileo's wording in the context of the previous note, where he speaks of "the essence of naturally accelerated motion" (*essentia motus naturaliter accelerati*, GG8: 197), indicates that he was claiming only to have uncovered the essential character of the motion, being aware of departures from it that would be attributable to accidental causes. Note also his reference to the "natural experiments" (*naturalia experimenta*) on which his reasoning is based, which would seem to suggest the free fall initiated in the table-top experiments rather than the acceleration observed on an artifact such as that down an inclined plane.

GALILEO GALILEI

TREATISE ON FOREKNOWLEDGES AND FOREKNOWNS

MS 27, Fols. 4r – 13r

Translated, with Notes, by William A. Wallace
From the Latin Edition of W. F. Edwards and W. A. Wallace

[ON FOREKNOWLEDGES AND FOREKNOWNS
IN GENERAL]

ON FOREKNOWLEDGES AND FOREKNOWNS IN PARTICULAR

[F2] *Second Disputation[1]: On Foreknowledges of Principles*

[F2.1] *First Question: Whether for every principle[2] [the answer to the question "Is it [true]?"[3] must be previously known[4]?*

[1] It seems not, [a] because one can have perfect knowledge of a conclusion from principles that are proper and immediate[5] and thus without any knowledge of first principles[6]; therefore [all principles need not be foreknown]. Also, [b] because special sciences[7] do not know the existence of first principles beforehand and nonetheless they have perfect knowledge of their conclusions; therefore knowledge of [the truth of] first principles is not required.

[2] First conclusion: first principles must be foreknown in some way[8] if the conclusion is to be perfectly understood, because the conclusion cannot be perfectly known unless all the principles regulating it and on which it depends in some way are known; but the conclusion depends in some way on first principles; therefore [first principles must be foreknown].

[3] Second conclusion: proximate and immediate[9] principles must be previously known if the conclusion is to be known, for Aristotle says in the first book of the *Posterior Analytics*, chap. 2, "that on account of which something is so must be even more so itself"[10]; but the conclusion is foreknown on account of the principles; therefore the principles themselves must be even more known.

[4] You ask: in what way are first principles known? The reply: in as many ways as there are different [kinds of] principles. For some are primary and most universal, and these are grasped solely through a knowledge of their terms, as is the principle "The whole is greater than its part." Others are known solely through sense knowledge, as is this: "Fire is hot." Some are known by induction, division, and hypothetical syllogism,[11] as is the principle "All instruction [given or received by way

of argument arises from preexistent knowledge"]. Yet others are known from experience, of which kind are medical principles, for example, that contraries are cured by contraries. And still others are known solely from custom, as are those of moral science, for we cannot understand these unless we have had practice with them ourselves.

[5] To the arguments: the reply to the first [1a] is apparent from what was stated in the first conclusion [2].

[6] To the second [1b]: I reply first that special sciences are not said to foreknow such principles, not that knowledge of them is not required, but because they are presupposed as obvious in such sciences. Add to this that anyone learning these sciences must be so disposed that once presented with principles that are obvious he assents to them immediately. I reply, second, that first principles in general pertain to metaphysics but when applied to this or that matter they pertain to special sciences.

[F2.2] *Second Question: Must nominal definitions*[1] *of the terms occurring in first principles be foreknown?*

[1] The reason for raising this question is that Aristotle says in text 2[2] that only the existence[3] of principles must be foreknown; but, if nominal definitions of the terms occuring in them are not previously understood, they will not be recognized as true; therefore, apart from the knowledge of existence, foreknowledge of nominal definitions is also required.

[2] Note: complex principles[4] cannot be given real definitions, from Aristotle in the seventh book of the *Metaphysics*, texts 13, 14, and 15[5]; the reason is that they cannot be assigned a genus and a proper differentia. But with regard to nominal definitions they can be defined in two ways: first, if some name is designated in the principle, then the meaning of that term is thereby defined, as in the example, "The *Illiad* is Homer's poems about the Trojan war"; in a second way, if the terms of a complex principle are explicated, then their meanings taken separately and conjointly are thereby made clear.

[3] This understood, I state that one must previously know not only the existence of principles but also the nominal definitions of their terms. This is proved from Aristotle, first book of the *Posterior Analytics*, texts 5 and 14,[6] where he expressly teaches this. Second, from reason: for knowledge of existence presupposes knowledge of the meanings of the terms; therefore if we are to know that first principles are true we must

previously know what their terms signify. Furthermore, this is the teaching on text 14 of Averroes, Philoponus, and Themistius.[7]

[4] You say, first: why then does Aristotle make no mention[8] of this foreknowledge? I reply, first: Aristotle intended to teach only the foreknowledges that immediately and proximately assist in understanding a demonstration, and foreknowledge of the meanings of terms is not of this type. I reply, second: Aristotle intended here to treat only two foreknowledges, the existence and meaning [of the subject], to which all others can be reduced; and thus there is nothing remarkable in his not being mindful of this foreknowledge.[9] Add, moreover, that Aristotle had already mentioned foreknowledge of the nominal definition in texts 5 and 14.

[5] You say, second: what should be said about primary and most universal principles? I reply: in such principles foreknowledge of nominal definitions is not necessary, first because primary principles can be grasped and understood without such foreknowledge, as is apparent in inventive science[10]; again, because directing or acting foreknowledge[11] are not needed for grasping primary principles.

[F2.3] *Third Question: Must principles be foreknown actually or habitually*[1]?

[1] First conclusion: the proper principles of a demonstrative science must be foreknown actually, both because Aristotle teaches this in the first book of the *Posterior Analytics*, texts 5 and 16[2]; in the second book of the *Posterior Analytics*, last chapter; and in the sixth book of the *Ethics*, chapter 3; where Aristotle teaches that a conclusion cannot be understood if its principles are not foreknown; and because such principles are the efficient cause of the science[3]; therefore there cannot be actual knowledge of a science if previous knowledge of those principles is lacking. Proof of the consequence: because cause and effect are correlatives[4]; but correlatives are such that one cannot be actually understood unless the other is also; therefore [principles must be foreknown actually].

[2] Second conclusion: axioms that enter {actually}[5] into an imperfect demonstration,[6] such as a reduction to the impossible, must be foreknown actually. This is proved from Aristotle, text 16; from Philoponus, on text 2; from Themistius, passim, on chapter 12. Second, for the same reason as given above, for axioms that enter into any demonstration are like its proper principles.

[3] Third conclusion: axioms that enter a demonstration neither actually nor virtually[7] need be known only habitually. The proof is from Aristotle, sixth book of the *Ethics*, chapter [3]; and because principles on which the conclusion itself is intrinsically dependent must be foreknown actually; but the conclusion is not intrinsically dependent on axioms, since these do not enter into a demonstration either actually or virtually; therefore [axioms need be known only habitually].

[4] You say: if the conclusion does not depend in any way on axioms, why must they be foreknown habitually? I reply, first: although the conclusion does not depend on them in the order of being,[8] it does depend on them in the order of knowing in a qualified way. Second, that we may be able to convince the obstinate.[9]

[F2.4] *Fourth Question: Should the principles of sciences be so evident[1] that they cannot be proved by any reasoning[2]?*

[1] It seems so, first, from Aristotle, text 15,[3] [a] because principles are supposed,[4] not proved; again, [b] because it is the task of the metaphysician[5] to prove principles, therefore not that of the other sciences; again, [c] because principles must be foreknown prior to any demonstration, therefore they are not proved. Avicenna, followed by some moderns,[6] feel that it pertains to the metaphysician to prove principles; for if Aristotle sometimes proved them in physics, he proved them as a metaphysician and not as a physicist.[7] The basis: because existence follows essence in common[8] precisely as it is essence; but essence in common pertains to the metaphysician[9]; therefore so does existence; therefore to prove it pertains to the metaphysician.

[2] Certain moderns distinguish two kinds of principle: some are of the object or in the order of being, others are of knowledge or in the order of knowing. They teach that principles in the order of being can be demonstrated *a priori*[10] from principles in the order of knowing. But quite the contrary: for principles in the order of being can be demonstrated only {*a posteriori*}[11] from principles in the order of knowing. Averroes, in the first book of the *Posterior Analytics*, com. 22, and in the seventh book of the *Metaphysics*, whom practically everyone follows, states that principles in the order of being, when unknown, can be demonstrated *a posteriori*.

[3] I say, first, principles in the order of being, when unknown, can be demonstrated *a posteriori*. This is proved, first, from Aristotle himself,

who in the first book of the *Physics* proved that protomatter[12] exists from transmutation, and who, in the eighth book of the *Physics*, inferred that a first mover[13] exists from the eternity of motion. Again, because principles that are unknown can be proved through effects that are more known in the science; but sometimes effects that are more known than first principles are available in it; therefore [principles can sometimes be proved through such effects]. Confirmation: for otherwise it would follow that the question of existence would be eliminated from all sciences except metaphysics.

[4] I say, second: principles that are not opaque to understanding, which are for the most part those in the order of knowing, are usually not proved in the sciences but are manifested[14] through some slight induction, or division, or hypothetical syllogism.

[5] I say, third, first and immediate principles[15] cannot be proved in any way, for otherwise they would not be first, since there would be others prior to them through which they would be proved. You say: what is to be said when first principles are opaque to understanding and cannot be manifested *a posteriori*? I reply: it pertains to a subalternating science[16] to prove such principles when they are proper, to dialectics[17] when they are probable, and to metaphysics when they are common.

[6] To the first objection [1a] I reply, with Philoponus and Averroes, that Aristotle is speaking of first and immediate principles, and these cannot be proved.

[7] To the second [1b] I reply by denying the antecedent, for although the existence of a thing in common, being associated with essence in common, pertains to the metaphysician, existence in particular, which is associated with this or that being in particular, is the concern of the particular sciences.

[8] To the third [1c] I reply that in demonstration *a priori* principles are foreknown, whereas in demonstration *a posteriori* they are [not foreknown but] sought.[18]

[F3] *Third Disputation: On Foreknowledges* {*of the Subject*}[1]

[F3.1] *First Question: What does Aristotle[2] mean by the term "is"[3] when he says that the "is" of the subject[4] must be foreknown?*

[1] Note, first: subject[5] can be taken in many ways; at present we take it to mean that of which some properties[6] or other are demonstrated, and

when taken in common it pertains to metaphysics,[7] when taken in particular for this or that thing it pertains to particular sciences.[8]

[2] Note, second: "is" is twofold, one applied to essence, the other to existence. The "is" of existence[9] is also twofold, either actual or potential[10].

[3] Note, third: authorities[11] on this matter agree on two points. First, [knowledge of] the actual existence of a thing is necessary for acquiring a science at the start[12]; the reason for this is that all our knowledge takes its origin from sense, which is concerned only with existence, and so [the actual "is" of the subject must be foreknown at the level of sense]. Second, in the development of a science[13] the "is" of the subject must be foreknown. They disagree, however, on how to take the meaning of this second "is"[14] of which Aristotle speaks.

[4] The first opinion is that of certain authors[15] who say that the "is" of the essence[16] of the subject must be foreknown, because a thing is related to knowing as it is to being, following Aristotle, and a thing must first have an essence before it has existence,[17] therefore [the "is" of its essence must be foreknown]; also, because no properties can be demonstrated of a subject unless it has a determinate quiddity[18]; therefore [knowledge of its essence precedes that of its existence].

[5] The second opinion is that of others[19] who hold that the potential "is" of the existence of the subject must be foreknown, but not the actual "is" of the existence. Their basis is this: because sciences demonstrate only that properties can exist in their subject, and for this only a potential existence need be foreknown, since that suffices.

[6] The third opinion is that of Averroes in the first book of the *Posterior Analytics*, comment 2; of Porphyry; of Cajetan on the first chapter of the first book of the *Posterior Analytics*, second question; and of many moderns[20] who say that the actual "is" of the existence[21] of the subject must be foreknown.

[7] Observe that three queries make this question difficult to treat: first, does the actual "is" of the existence of the subject have to be foreknown, in view of the fact that many things are known by us at all times even though they do not exist at all times[22]; second, why should it not be sufficient to foreknow solely the "is" of the {essence}[23] of the subject; and third, why does it happen that in some demonstrations it is not necessary to foreknow the existence of the subject? I will answer these difficulties in the questions that follow.

[8] First conclusion: the "is" of the {essence}[24] of the subject must be

foreknown, for if it is not, there cannot be a demonstration.[25] The proof
of this is from the arguments favoring the first opinion [4].

[9] Second conclusion: in real sciences[26] the actual "is" of the existence
of the subject of demonstration must be foreknown, at least for its places
and times and removing its impediments,[27] in cases where either a
property or some other predicate is shown to inhere in it. The proof of this
conclusion: human sciences[28] are concerned with existents, and so the
existence of their objects must be foreknown, lest they prove to be
fictitious. Second, from Porphyry: such existence of the subject must be
foreknown by reason of the fact that it underlies its proper attributes; but
an actual {"is"}[29] of existence underlies its proper attributes; therefore
[such an actual "is" of existence must be foreknown]. Third, the question
of existence is the first of all [scientific questions],[30] and it seeks the actual
existence of a thing; therefore the [answer to the question] "Is it?" must
be foreknown before [the answer to] the question "what kind is it?"
Fourth and last: otherwise it would follow, contrary to Aristotle, that no
more could be known of the subject than of the property, for anyone who
wishes to prove a property of a subject presupposes that the property can
be present in it.

[10] Third conclusion: in sciences the actual "is" of the existence of the
subject need not always be foreknown. This is apparent in demonstrations
that show that existence pertains to their subject, as is to be seen in
demonstrations in which protomatter is shown to exist through
substantial change or in which a first mover is argued from the eternity of
motion.[31] You say: why then did Aristotle state that the existence of the
subject must be foreknown? I reply: Aristotle was speaking of the subject
of demonstration in its proper sense[32]; second, Aristotle was speaking in
an indeterminate way, in the sense that the existence of some subjects [of
demonstration] must be foreknown, but not of all.

[11] Fourth and last conclusion: in rational sciences, if there be such,[33]
since many regard logic as of this kind, the actual "is" of the existence of
the subject need not be foreknown, but only its "is" as an object[34] [of the
intellect]; for the type of existence that must be foreknown is that
appropriate to the subject, and the existence of the subject of a rational
discipline is an existence in the intellect; therefore [the actual "is" of its
existence need not be foreknown].

[12] The first objection from the second opinion [5]: sciences abstract
from existence,[35] and therefore they cannot foreknow the existence of
their subjects. I reply: if we look at the formal consideration of the

sciences, they do indeed abstract from the existence of their subjects, for they consider universals, and universals cannot be known as existents. If, however, we attend to the condition *sine qua non* [i.e., if the subjects do not exist they cannot be known in the sciences], I deny that they abstract from existence.

[13] Second objection: for a thing to exist is a contingent matter; but sciences are not concerned with contingent matters[36]; therefore [sciences are not concerned with the existence of their subjects]. I reply: there are two kinds of existence, the particular existence of this or that individual, and that is contingent; and the existence of the species, and this is considered in the science, and it is necessary, on the supposition of there being a universe,[37] at least for its temporal duration.

[14] Third objection: existence is proper to individuals; but science does not consider individuals[38]; therefore it does not consider existence. I reply: this or that existence in particular is proper to individuals, but the existence that follows on a universal nature, not in the abstract but indeterminately in some individual, is proper to the species. [. . .][39] Note here, however, that the existence of the species is more intended by nature than is that of individuals, because nature is perfected more by species than it is by individuals.

[15] Fourth objection: if God were to destroy all corporeal species, there would still be science of them, and yet their existence could not be presupposed; therefore [their existence need not be foreknown]. I reply: granted this event, there would still be science, but then it would be necessary to foreknow the time of existence of the subjects, removing the impediment of the divine will,[40] and that would suffice.

[16] Fifth objection: even if God created nothing and if he never decreed to create anything, angels[41] could still know some of the properties that are proper to corporeal nature; therefore [actual existence is not necessary for scientific knowledge]. I reply, first: we are speaking here of human sciences.[42] Second, I reply that then angels would know that those properties are possibly present in such subjects, not that they are there actually, in the same way that their nature is also possible.

[17] Sixth objection: sciences show merely that properties are possibly present[43] in their subjects; therefore [their concern is only with possible existence]. Confirmation: because sciences prove only that certain propositions are necessary; but propositions of this kind abstract from the "is" of existence; therefore sciences also [abstract from the "is" of existence]. I reply: it may be that sciences show only that properties are

possibly present in subjects; yet, for them to show that properties are possibly present in real subjects, they must foreknow the "is" of the existence of such subjects. I reply, second: generally sciences prove that properties are present actually in their subjects, [therefore generally they are not concerned solely with possible existence]. To the confirmation: I reply that a proposition can be taken in two ways, either considering the connection of its terms, and so considered it abstracts from all existence; or considering how it states that a predicate is present in a subject, and so considered I deny that it abstracts from existence.

[18] Final objection: mathematics, abstracting as it does from being and goodness,[44] does not presuppose the "is" of existence of its subject. I reply: mathematics abstracts from all existence when demonstrating properties, yet it foreknows the existence of its subject, because, since it is a human science,[45] it is concerned with existence.

[F3.2] *Second Question. Can a science demonstrate the "is" of existence of its adequate subject[1]?*

[1] The first opinion is that of Scotus in the first book of the *Sentences*, third question, and in the first book of the *Metaphysics*, first question; Antonius Andreas follows him there, and so do all Scotists. This opinion holds that a science can demonstrate the existence of its total subject only *a posteriori*.[2] [a] First: because sometimes an effect can occur that is more known than the total object[3] itself; therefore [the effect can be used to demonstrate the existence of the total object]. Proof of the antecedent: because in physics change is more known than a natural body; therefore [change can be used to demonstrate the existence of a natural body, which is the total subject of physics]. [b] Second: just as a partial science[4] stands in relation to its partial object, so a total science[5] stands in relation to its total object; but a partial science can demonstrate the existence of its subject[6]; therefore a total science can do the same for its [subject]. The minor is proved by Aristotle himself, who, in the first book of the *Sophistical Refutations*, first chapter, proves the existence of the subject of that book [i.e., the sophistical argument], and in the first book of the *Posterior Analytics*, first chapter, [proves] that a demonstration [i.e., the subject of the first book] exists, and in the third book of *De caelo* that fire [i.e., the subject of the third book] exists; therefore [a partial science can prove the existence of its partial subject]. [c] Third: because it pertains to the same science to consider the [answers to the questions] "Is it?" and

"What is it?"; but a total science answers the [question] "What is it?" of its subject[7]; therefore it can answer the [question] "Is it?" of it also. [d] Fourth and last: because, among all the questions usually asked about a subject the first is "Is it?," and a total science ought not to omit consideration of that [question]; therefore [a total science should prove the existence of its subject].

[2] The second opinion is that of Themistius, on the first book of the *Posterior Analytics*, chapter 21, and on the second book of the *Posterior Analytics*, chapter 11; of Averroes, on the first book of the *Posterior Analytics*, commentary 36, on the second book of the *Physics*, last commentary, and on the first book of the *Metaphysics*, commentary 23; and of Cajetan here, [i.e., on the first book of the *Posterior Analytics*], first question; of the Conciliator, and of others, all of whom say that a science cannot demonstrate the existence of its subject either *a priori* or *a posteriori*.

[3] Note that there are three kinds of things that are found in sciences: some of these are completely known, and these cannot be demonstrated since demonstration is needed only for the proof of what is unknown, and things that are directly apprehended do not require proof; others of them are not known, and these can be proved either *a priori* or at least *a posteriori*; and yet others are partly known and partly unknown. The last kind, though they cannot be demonstrated by any type of demonstration, nonetheless can be manifested[8] either by induction or by hypothetical syllogism.

[4] I say, first: the existence of the total subject cannot be demonstrated *a priori* in its own science, nor can it be proved *a posteriori*. I said "in its own science," for it can be proved in a higher science,[9] just as it can be manifested in its own science. Proof: first, from Aristotle, first book of the *Posterior Analytics*, 25, where Aristotle teaches that the existence of the total subject must be foreknown; therefore not sought; therefore not demonstrable. Second, things considered in a science are either principles or parts or properties of the object; but for all of these the existence of the subject itself is presupposed as most evident; therefore [the existence of the total subject must be foreknown]. Third, because science and its knowable object are correlatives[10]; but correlatives are coincident in nature, time, and knowledge; therefore, since the object of a total science is correlative with the science, it cannot be demonstrated by the science.

[5] I say, second: the existence of the principal subject[11] cannot be

proved by any species of demonstration within the science, because such a subject is the culmination of all the partial subjects, and because, since the total subject is especially dependent on the principal subject, the existence of the one cannot be known without that of the other.

[6] You ask for the reason why such subjects are known directly. I reply: because it is intrinsic to the formality of a subject that its existence be grasped directly, so that, if it were to be demonstrated, it would cease to be a subject, since a subject is the datum which, by universal consensus, underlies everything said in the science. Second, because nothing is more known in a science than that it be provided a subject.

[7] To the first objection [1a], I reply: for something to be demonstrated it is not sufficient that some effect be given that is more known; it is also required that what is to be demonstrated be not known.

[8] To the second [1b], I reply: in the first book of the *Sophistical Refutations*, first chapter, Aristotle does not prove that the subject of that work exists but merely points it out by giving examples; likewise, in the first book of the *Posterior Analytics* he does not prove that demonstration exists, as we have shown above. Second, I reply by denying the minor; for it pertains either to the total science or to some part of a higher science to demonstrate the existence of a partial subject.

[9] To the third [1c], I reply: here[12] Aristotle teaches only that the particular sciences do not consider the existence of their subjects because they do not consider their quiddity.[13] How this text[14] is to be understood we shall see in the following question.

[10] To the fourth and last [1d], I reply: [the answer to] the question "Is it?" with respect to the total subject is known for the reasons given in the first conclusion. With respect to that of the partial subjects, it is not to the point. You say: at the beginning of a science it is customary to inquire about the object of that science; therefore [its object can be demonstrated]. I reply: at the beginning of a science it is customary to inquire only about what unifies the various properties of the science. And if one were to manifest the existence of the total subject, one would do this as having the habitus[15] of a higher science.[16]

[F3.4] *Fourth Question*[1]: *Can a science demonstrate the [answer to the question] "Is it?" of its partial subject?*

[1] Some[2] deny this, both [a] because to prove existence pertains to the metaphysician, and [b] because the total object is nothing more than all

the partial objects taken together; therefore, if one has doubts about the existence of one partial object he will also have doubts about the total object as a whole.

[2] Others[3] affirm it, both because it pertains to the same science to consider the quiddity and the existence of its object, and because in the third book of *De caelo* Aristotle proves that fire exists, and in the second book of *De anima* that an agent sense exists in us, and, in the third book, that an agent intellect exists. You say: Aristotle did this as one having the habitus of a higher science.[4] But, to the contrary: first, because it is not permissible to pass from one genus to another[5]; second, because the middle terms through which Aristotle proved the existence of such subjects were physical; therefore [they were not metaphysical, and thus he did not prove their existence through the habitus of a higher science].

[3] Note, first: the question has two senses. The first is whether a science can demonstrate the existence of its partial subjects *a priori*. The second is whether it can do so *a posteriori*.

[4] Note, second: for a science to show the existence of its partial subjects *a priori* two things are required. One is that the quiddity of the subject itself be known, for just as the "is" of the subject depends on the quiddity, so also does the existence, which is the actuality of the subject,[6] depend on the same quiddity; for this reason the existence cannot be demonstrated perfectly unless the quiddity of the subject is previously known. The other is that the existence to be demonstrated must not be known, for if it is known it cannot be demonstrated. But for the existence to be demonstrated *a posteriori* only the second point is required, namely, that the existence not be known.

[5] These matters presupposed, I say first: the existence of a partial subject can be demonstrated within a science *a posteriori* when it is not known, but this rarely happens, as the second opinion [2] maintains. The proof, first: it can sometimes happen that the existence of a subject not be known in a science, and yet that some effects be more known in it through which this existence can be demonstrated; therefore [the existence of a partial subject can sometimes be demonstrated in a science *a posteriori*]. Second, otherwise the question "Is it?" would be [pointless][7] in every science with the sole exception of metaphysics. Nor can you say it would not be pointless because the existence of principles can be manifested in the science, for the same reasoning applies to principles as to partial subjects; therefore, if the existence of partial subjects cannot be demonstrated, neither can that of principles.

[6] I say, second: the existence of a partial subject cannot be demonstrated within a science *a priori*, not because the science does not consider the quiddity of its partial subjects, but because the quiddity of a subject cannot come to be known without knowledge of the existence of that same subject. I explain: because it is necessary for one wishing to know the quiddity of the subject to investigate the parts of its definition,[8] and then to recognize that such a quiddity exists; and, since the quiddity of the subject cannot be known to exist unless there is previous knowledge of the subject's existence, the quiddity will not be demonstrable.

[7] You ask: why is it that the existence of principles is frequently proved within a science, and that of partial subjects only rarely? I reply: because it is of the nature of a subject to be subjected and to be presupposed to other things, and so it should be most known. The same reasoning does not apply to principles.

[8] To the first objection [1a], I reply, first: it pertains to the metaphysician to prove existence in common but not in particular. I reply, second: it pertains to the metaphysician to prove through common principles and by demonstration that reduces to the impossible. You object: in the sixth book of the *Metaphysics*, first text,[9] Aristotle states that the particular sciences do not consider the quiddity of such subjects; therefore the response is not valid. I reply: Aristotle was speaking there of existence in common, which is a metaphysical predicate, just as he was speaking of quiddity in common, after which existence in common follows. And that this is true can be confirmed from this consideration, for otherwise Aristotle would have contradicted himself when he says that the consideration of all causes pertains to the particular sciences, among which the formal cause[10] is also included, and, as a consequence, the quiddity of the thing.

[9] To the second [1b], I reply: a science presupposes the existence of its total subject in a general way, not distinctly and perfectly. Thus it follows that the total subject can be given as known even though there can be doubts about a partial subject.

[F3.5] *Fifth Question: Can a science manifest the real definition*[1] *of its subject and explain*[2] *its* {*existence*}[3] *apodictically?*

[1] Two things are sought in this question, as is apparent from the title. Concerning the first, whether a science [can manifest a real definition of its subject], I say, first: a science can give real definitions of its partial

is included under foreknowledge of the fact, which applies to terms];
finally, because the definition of a term differs in many ways from the
definition of a thing's quiddity. For the quiddity of a thing is one, perfect,
and is said of simple [and]⁹ perfect entities; it never is "directing"
foreknowledge but always is "acting"¹⁰; and it is known only to experts¹¹;
just the opposite properties, however, characterize the definition of a
term; therefore [the definition of a term is quite different from the
definition of a thing's quiddity].

[4] The solution to objections that can be brought [against this
teaching] is apparent from what has been said in the question "How many
things should be foreknown¹²?"

[F4] *Last Disputation: On Foreknowledges of the Property¹ and of the Conclusion²*

These two matters, though differing by nature from each other, will be
joined in a single disputation because we are going to make only a few
notations about them.

[F4.1] *First Question: Must the existence of a property³ be foreknown?*

[1] It seems that it must be: [a] for we always presuppose that the property
exists and then we seek the apodictic reason why it is present⁴ in a
particular subject; second, [b] because even before we demonstrate a
property to be in a particular subject and the apodictic reason why it is
there, we must ascribe it to some subject; therefore we presuppose its
existence.

[2] Note, first: this question is to be understood not only of a proper
attribute⁵ but also of all properties that are demonstrated of a particular
subject.

[3] Note, second: there are three kinds of demonstration, of the fact,⁶
of the reasoned fact,⁷ and most powerful.⁸ That of the fact demonstrates
[a cause] from an effect; that of the reasoned fact gives the reason why a
property exists in a subject; and that which is most powerful both gives the
reason why a particular property exists in a subject and proves the
existence of that property.

[4] Note, third: there are two kinds of property: one is convertible with
its subject,⁹ as risibility with respect to man; the other is not convertible,¹⁰
as white with respect to man.

[5] I say, first: the nominal definition[11] of a property must always be foreknown, for otherwise we could neither demonstrate the property's existence nor give the proper reason for this.

[6] I say, second: in a demonstration of the fact, the fact of the property's existence must always be foreknown. The reason: for otherwise its essence[12] could not be demonstrated by means of it.

[7] I say, third: in a most powerful demonstration, which proves why the property[13] is present in the subject and also the property's existence, the existence of the property is never foreknown. The reason: anything being proved is not foreknown; therefore, since the existence of the property is being proved in such a demonstration, it must not be foreknown.

[8] I say, fourth: in a most powerful demonstration, which proves why the property is present in the subject and also the property's existence (I am speaking now of a property that is not convertible, for the preceding conclusion is to be understood of a convertible[14] property), it is not necessary that the property's existence be foreknown, although absolutely and simply speaking[15] there is nothing to prevent its being foreknown. Note, however, that the existence of a property that is to be demonstrated can never be foreknown, for what is foreknown cannot be proved, and because a demonstration proves only what is not known.

[9] I say, fifth and last: in a demonstration that proves only why a property is present in a subject, it is never the case that the nominal definition and the existence of the property not[16] be foreknown. The reason: for in such a demonstration the existence of the property is not proved and it is posited to exist; therefore [its nominal definition and existence must be foreknown].

[10] You ask: if this is so, why is it that Aristotle makes no mention[17] of the foreknowledge of the existence of such a property? I reply, first: because Aristotle's intention, as is obvious from the foregoing, was merely to enumerate the two general kinds of forekowledge,[18] not all ways in which things can be foreknown; therefore [he does not mention this particular type of foreknowledge]. I reply, second: Aristotle offered the example of most powerful demonstration, and in this species of demonstration the existence of the property is never foreknown.

[11] To the first objection [1a]: the reply is obvious from the foregoing.

[12] To the second [1b], I reply: in the premises [of a demonstration] one knows only the connection of an attribute with a subject, not the attribute's existence. I reply, second: in the premises the existence of the

property is known by the "acting" knowledge of the premises, whereas here the question concerns "directing" knowledge;[19] therefore [the existence of the property is not relevant here].

[13] One may ask: must the real definition[20] of a property sometimes be foreknown? I reply: if we are speaking of "directing" foreknowledge, the real definition of a property is never foreknown; but if we are speaking of "acting" foreknowledge, I reply in the affirmative, for in this case knowledge of the real definition is reducible to the "acting" knowledge of the premises.

[F4.2] *Second Question: Is the conclusion known at the same time[1] and with the same priority[2] as the premises?*

[1] Note, first: one thing is said to be prior to another in two ways, either in time, or by nature; in time, morning is prior to afternoon; by nature, the cause is prior to the effect, as fire before heat and the sun before its light.

[2] Note, second: major and minor premises[3] can be considered in four ways. In the first way, as they are particular propositions that either affirm or deny something of something else; in the second way, as they imply a relationship to the conclusion as cause to effect[4]; in the third way, as they are the reason, the means, and {the path}[5] to knowing the conclusion; in the fourth way, as they have a syllogistic structure following the proper mode and figure.

[3] I say, first: the major and the minor, if taken in the first way, are known before the conclusion by a priority of nature and of time. The proof for the first part: because things that are prior by nature are known first; but propositions in a syllogism are prior by nature to the conclusion; therefore [propositions in a syllogism are known before the conclusion]. For the second part, the proof: as things are related in the order of being, so they are related in the order of knowing[6]; but propositions precede the conclusion in time; therefore [they precede the conclusion in time of being known]. Confirmation: because they can be known and understood even when the conclusion is not.

[4] I say, second: the major and the minor, if taken in the second way, are known not only simultaneously by nature and in time, but also by the same act.[7] Proof of the conclusion with respect to the first part: for things that are simultaneous by nature and in time are known simultaneously by nature and in time; but premises considered in this way are simultaneous

with the conclusion by nature and in time; therefore [they are known simultaneously with the conclusion by nature and in time]. Proof for the second part: for the dependence of one correlative is of the essence of the other; but things that have one and the same essence are known by the same act; therefore [the premises are known by the same act as the conclusion].

[5] I say, third[8]: the major and the minor, if taken in the third way, are known by the same act, in such a way, however, that the act of knowing bears first on the premises as on the means and then on the conclusion as on the end. I explain the conclusion: since the premises are like the path or means to the conclusion and the conclusion is like the end of the premises, the premises cannot be known formally as a path unless the conclusion is known as its end, and vice versa; but because the path is prior by nature to its end, the act of knowing must first bear on the premises as on the path before it bears on the conclusion as its end. Confirmation, using the example of vision: in the same act sight focuses on light and on color, but on light as a means [of seeing] that is prior by nature, then on color as the terminus [of sight]. Proof from reason: we are brought by one and the same act to the means and to the end; but the premises are the means, the conclusion the end; therefore they are known in the same act.

[6] First objection: the premises and their knowledge are the cause of knowledge of the conclusion; therefore they cannot be known in one and the same act. I reply: the premises taken by themselves are the cause of the conclusion and are not known at the same time as it is; considered, however, as they are the path and the means to the conclusion they provide the formal reason for grasping the conclusion.

[7] Second objection: the many precisely as many cannot be known in the same act; but the premises and the conclusion are many; therefore [they cannot be known in the same act]. I reply: the premises and the conclusion, when known in the same act, are known as one, namely, insofar as the premises are the means and the formal reason for knowing the conclusion, and insofar as they are united with their end in the conclusion itself.

[8] Third objection: it would follow that there would be both {assent}[9] and dissent in one and the same act, if this solution were true; but the consequent is absurd; therefore, so is the antecedent. Proof of the major: let there be a syllogism with one of its premises {affirmative} and the conclusion {negative}[10]; if one were to assent to that syllogism one would

be assenting and dissenting at the same time; for one would be assenting to the affirmative premise and dissenting from the negative conclusion. I reply: in such a case there is only a single assent, which bears on the premises as on the means and on the conclusion as on the end; nor is it true, as some have falsely believed, that any knowing of a negative proposition is a dissent, for it truly is an assent.

[9] I say, fourth and finally: the premises taken in the fourth way are known at the same time as the conclusion. I explain the conclusion: as soon as one recognizes that the premises in a particular demonstration have the proper syllogistic structure, one assents to its conclusion immediately, in such a way, however, that one assents first to the major premise, {then}[11] to the minor, and lastly to the conclusion, but with no time delay in between; and this is what I understand conclusion to mean when I say that the conclusion is known at the same time as the premises. Note, however, that in a demonstration there is a threefold assent which does not pertain to our subject, namely, that of the major, that of the minor, and that of the conclusion, each one of which comes before the other. Proof of the conclusion, from experience: if one knows that all fire is hot, and then [later] that this is fire, one knows immediately and {at the same time}[12] that this fire is hot. Proof from reason: every natural cause that is sufficient to produce its effect does so necessarily as soon as the requisite conditions are provided[13]; but the intellect, together with the knowledge of principles, is a cause that is natural and sufficient to produce science; therefore [it grasps a conclusion as soon as it understands the premises on which it is based].

[10] You ask, first: whence does it come about that, when we understand the premises, we must assent to the conclusion necessarily? I reply, with St. Thomas in the first book of the *Perihermenias*: because there are certain things that follow necessarily from the knowledge of principles; for this reason, when the latter are known, the conclusions that are contained in them virtually and are apt to be inferred are known necessarily.

[11] You ask, second: in what genus [of cause] does knowledge of the conclusion depend on knowledge of the premises? I reply: knowledge of the conclusion, considered as it comes to be from the terms of the premises, depends on knowledge of the premises in the genus of material cause, from Aristotle, fifth book of the *Metaphysics*, chapter 2; but knowledge of the conclusion as it is inferred from the premises depends [on them] in the genus of efficient cause.

[12] Objection: a natural cause that is sufficient to produce its effect, given all the requisite conditions, does [not][14] act necessarily. I reply by denying the antecedent, because, even though this is true with regard to man, since he is free, it is not true of other things that are determined to a single way of acting.

NOTES AND COMMENTARY

F2: *On Foreknowledges of Principles*

F2.1: Must every principle be foreknown to be true? *For background, see Secs. 4.1 and 4.6 of* Galileo's Logic of Discovery and Proof, *for applications, Secs. 5.1 and 6.5b*

[1] *Second Disputation*: Galileo's labeling this the "second disputation" is an indication that another disputation had preceded it. The title of the missing disputation is unknown, since Carbone reorganized Vallius's notes and did not preserve the titles of the disputations into which the treatise was originally divided. From a study of Vallius-Carbone's version one can surmise that the first disputation was entitled "How many foreknowledges and foreknowns are there?" It is probable that Galileo wrote out the disputation and that the folios of the codex containing it were subsequently lost. Less likely is the possibility that Galileo did not appropriate it and began his treatment directly with the second disputation. The entire content of this second disputation, like that of the other three in this treatise, has counterparts in Vallius-Carbone. When composing it Galileo apparently appropriated about 25 % of the material available in his source; for this type of information and relevant word counts, see Lat. Ed. xxxiv–xxxv.

[2] *for every principle*: that is, for every premise or statement that enters into the demonstration or bears on its conclusion, either directly or indirectly.

[3] *"Is it [true]"*: Lat. *an sit*, literally "Is it?" As applied to a principle this is equivalent to asking whether the principle is true and known to be such – an instance of complex truth; see D2.1.2–3.

[4] *known beforehand*: that is, foreknown, or known before the demonstration can be understood and assent given to its conclusion.

[5] *proper and immediate*: that is, principles that are proper to a particular subject matter and that elicit immediate assent. An example would be definitions, such as that of a triangle; see D2.5.

[6] *first principles*: axioms or common principles that underlie all reasoning, examples of which are given in paragraph [4]. Sometimes "first principles" refers to proper principles, but in this context the emphasis is on general or common principles; see D2.5.4–5.

[7] *special sciences*: Lat. *scientiae particulares*, sciences concerned with particular subject matters, also known as "partial sciences" and thus opposed to "total sciences," those broader in scope; see F3.1.1 n. 8, and F3.2.1 n. 4. In this context, as can be seen from the reply in paragraph [6], the special sciences are opposed to metaphysics, the science of being in general.

⁸ *in some way*: Lat. *aliquo modo*, not absolutely or simply but in a qualified way. Various possible qualifications, such as being foreknown actually, habitually, or virtually, are discussed in F2.3.

⁹ *proximate and immediate*: the same sense as "proper and immediate" in paragraph [1]; see n. 5 above.

¹⁰ "*that on account...more so itself*": an expression occurring at 72a29 that is difficult to translate but is usually rendered into Latin as *propter quod unumquodque tale et illud magis*; it also occurs in Aristotle's *Metaphysics* at 993b24. The sense of the axiom, much used in scholastic reasoning, is this: if water is made hot by fire, then the fire must possess heat to a higher degree than the water. Galileo uses the axiom again in D2.6.10.

¹¹ *induction, division, and hypothetical syllogism*: this expression occurs again at F2.4.4 and, in truncated form, at F3.2.3. The principle it is invoked to support is, in effect, the opening sentence of the *Posterior Analytics* on which the entire treatise on foreknowledge is based, namely, "All teaching and all learning through discourse arise from previous knowledge" (71a1). In the remaining sentences of his first paragraph, Aristotle goes on to offer a complete "induction" that is based on a "division" of the ways in which all learned disciplines are acquired. He mentions "syllogism" in this context, but not "hypothetical syllogism." Perhaps the expression is to be understood in this way: to grasp the truth of this principle, all one need to do is divide knowledge acquired by teaching into its various kinds (division), examine the different cases to show how some other knowledge is presupposed in each kind (induction), and then argue hypothetically, if this is true of each and every kind, it must be true of all (hypothetical syllogism). See D3.1 n. 38, Sec. 4.6 of *Galileo's Logic of Discovery and Proof*.

F2.2: Must nominal definitions of terms in first principles be foreknown? *For background, see Secs. 4.1 and 4.2 of* Galileo's Logic of Discovery and Proof, *for an application, Sec. 6.8*

¹ *nominal definitions*: Lat. *quid nominis*, or meaning of the term, as opposed to *quid rei*, or meaning of the thing. The first is usually called a nominal definition, the second a real definition.

² *in text 2*: medieval and Renaissance commentators on Aristotle divided his exposition into sections, frequently of paragraph length, which they numbered for purposes of ready reference. Here the reference is to the section marked off approximately by the Bekker numbers 71a12–17.

³ *existence*: Lat. *an sit*; see F2.1 n. 3.

⁴ *complex principles*: Lat. *principia complexa*, meaning propositions composed of subject and predicate that serve as principles of knowing, as opposed to entities that serve as principles of being for other entities (e.g., elements in relation to compounds) and in this sense are simple. Complex principles are known by complex truth, simple principles by simple truth; see D2.1.2. For the difference between principles of knowing and principles of being, see F2.4.2–4.

⁵ *texts 13, 14, and 15*: that is, *Metaphysics* Z.4, 1030a3–b5.

⁶ *texts 5 and 14*: text 5 corresponds to the entire second chapter of the book, 71b10–72b4, in which Aristotle sets out the requirements for knowing any subject scientifically; text 14,

to the latter part of the fifth chapter, 74a33–b4, where he explains what it means to know universally and without qualification.

[7] *Averroes, Philoponus, and Themistius*: commentators on the text of Aristotle. For biographical and bibliographical details on these authors and their works, see the alphabetical listing by author in the Biographical and Bibliographical Register at the end of the volume.

[8] *make no mention*: the question, as stated, is withdrawn in the last sentence of this paragraph, where it is claimed that Aristotle *does* make mention of this foreknowledge in texts 5 and 14, in chapters 2 and 5 respectively. Perhaps its sense is that Aristotle makes no mention of it in chapter 1, where he expressly enumerates only two foreknowledges, existence and meaning (71a11–14), as stated in the second reply to the query.

[9] *this foreknowledge*: that is, the meaning of terms in principles, not the meaning of the subject; see F3.6 n. 2 and F3.6.4 n. 12.

[10] *inventive science*: alternatively, investigative science or dialectics, which employs topical reasoning whose methodological procedures are set forth in the *Topics*. It is characteristic of such science to proceed from common opinion, that is, from principles that are accepted by all and thus do no require special foreknowledge to be understood. See Sec. 2.8 of *Galileo's Logic of Discovery and Proof*.

[11] *directing or acting foreknowledge*: Lat. *praecognitio dirigens...agens*, technical terms that are explained in the missing "first disputation" of this treatise and are here presupposed; they occur again at F3.6.3 and F4.1.12–13. The distinction is attributed to Averroes. As Vallius explains it, directing foreknowledge assists one in knowing but does not make one know, whereas acting foreknowledge actually produces new knowledge. Vallius-Carbone liken directing foreknowledge to a *conditio sine qua non* and acting foreknowledge to an efficient cause in the learning process. (For a fuller explanation, see Sec. 4.1 of *Galileo's Logic of Discovery and Proof*.) The most universal principles involved in demonstration do not require this assistance nor are they efficient causes in a proper sense; thus they do not require either type of foreknowledge.

F2.3: Must principles be foreknown actually or habitually? *For background, see Secs. 3.2 and 4.2 of* Galileo's Logic of Discovery and Proof, *for applications, Secs. 5.3 and 6.2a*

[1] *actually or habitually*: to have foreknowledge actually would be to have explicit knowledge of a principle, whereas to have it habitually would be to be able to reason with it without being explicitly aware of it. Thus the sense of the question is this: in working a problem in arithmetic does one have to have explicit knowledge of an axiom such as "equals added to equals the results are equal," or does the ability to use the axiom suffice?

[2] *texts 5 and 16*: for text 5, see F2.2.3 n. 5; text 16 is at 74b13–26.

[3] *efficient cause of the science*: in the sense of being the efficient cause of the conclusion; see D2.6.14. For a fuller explanation of the types of causality involved in demonstration and in science, see D1.1.2.

[4] *cause and effect are correlatives*: that is, a cause is a cause strictly speaking only when it is actually producing an effect, and thus the one, either cause or effect, cannot be known without the other; further use of this line of reasoning is made in Galileo's treatment of the demonstrative regress, D3.3.7–8.

⁵ *enter {actually}...foreknown actually*: in appropriating this conclusion Galileo inadvertently left out the first "actually" and thus obscured its meaning. If an axiom states an impossibility, for example, and this is one of the premises on which the demonstration is based, then the axiom must be foreknown actually for the conclusion to be understood.

⁶ *imperfect demonstration*: on the various types of demonstration, including demonstration to the impossible, see D1.1.1 and D3.1. The reason why demonstration to the impossible is said to be imperfect is that it argues from false premises rather than from premises that are true and certain (D3.1.4); yet it is true demonstration (D2.1.5) in that it leads to a necessary conclusion.

⁷ *virtually*: neither actually nor habitually, but in a way intermediate between the two. This distinction was invoked by the Dominican commentator on St. Thomas, Cardinal Cajetan, to explain how causes, and demonstrations based on them, could be applied in theology to explain the divine attributes. It was further used by commentators on the *Posterior Analytics* to solve problems relating to premises and the ways in which they can be said to be first, prior, and immediate; see D2.2.7 and D2.4.4.

⁸ *order of being...order of knowing*: in other words, though axioms may not serve as principles in the order of being they may nonetheless serve as principles in the order of knowing; this distinction is elaborated in the following question, particularly in F2.4.2–4.

⁹ *to convince the obstinate*: the statement here is cryptic, but the meaning seems to be that reduction to the impossible is a type of *ad hominem* argument that can be used to convince one of the absurdity of one's position, in which case it becomes a reduction to the absurd (*reductio ad absurdum*).

F2.4: Should principles be self-evident and incapable of proof? *For background, see Secs. 3.3, 4.2, and 4.7 of* Galileo's Logic of Discovery and Proof, *for applications, Secs. 5.1, 5.2, 6.2b, 6.4a, and 6.5b*

¹ *so evident*: Lat. *ita nota*, to be taken in the sense of the principles being *per se nota* or self-evident and thus not requiring proof through antecedents that are more known. See Sec. 4.7 of *Galileo's Logic of Discovery and Proof*.

² *cannot be proved by any reasoning*: Lat. *nulla ratione probari possint*, that they cannot be established by any type of discourse.

³ *text 15*: in ch. 6, 74b5–13.

⁴ *supposed*: Lat. *supponuntur*, a technical term used by Aristotle to indicate that principles in a demonstration function as "suppositions," i.e., as a type of immediate proposition. Unlike axioms, the other type of immediate proposition, suppositions do not have to be foreknown; usually they are accepted by those to whom the demonstration is being proposed and on this account do not have to be proved. See D2.3.7; also Sec. 4.2 of *Galileo's Logic of Discovery and Proof*.

⁵ *the metaphysician*: in its expositive role, metaphysics is the science of being *qua* being, concerned with explicating the universal properties of being as such. As the supreme science it also takes on a critical role, justifying and defending principles (including first principles) against those who would reject them or attempt to deny them. See Sec. 3.3b of *Galileo's Logic of Discovery and Proof*.

⁶ *moderns*: Lat. *recentiores*, possibly referring to contemporary but unnamed

Neoplatonists. Galileo's source here speaks of "Avicenna and others," where the "others" refers to Plato, Plotinus, Themistius, and Simplicius (for details, see Lat. Ed. 129).

[7] *as a physicist*: that is, as a natural philosopher, equivalent in that day to a natural scientist. A number of principles in Aristotle's *Physics* have counterparts in his *Metaphysics*, whence the ground for the objection.

[8] *existence follows essence in common*: a statement with Neoplatonic overtones. Essence and existence are the two basic principles of all being, and in creatures, according to Aquinas, they are really distinct from each other. For a being to exist it must first have an essence; thus there is a certain priority in the principles themselves, with essence preceding existence in the ontological order.

[9] *the metaphysician*: that is, the person who studies being in common, the subject of metaphysics; see n. 5 above.

[10] *a priori*: from cause to effect, from something prior in the order of being to something posterior in the same order.

[11] {*a posteriori*}: omitted by Galileo here and supplied for sense; the opposite of *a priori*, i.e., from effect to cause, from something posterior in the order of being to something prior in the same order. (This reading emends that found in the Latin Edition, p. 6, whose critical apparatus for line 21 indicates that Galileo omitted *a priori* at this point and wrote *nisi* instead. If Galileo did so the text would read, in translation: "principles in the order of being cannot be demonstrated {*a priori*} through principles in the order of knowing." Although this conveys the correct sense, in view of the presence of the *nisi* it is more faithful to the text to add the *a posteriori* after the *nisi* than to delete the *nisi* and substitute *a priori* instead.)

[12] *protomatter*: Lat. *materia prima*, Gr. *hylē protē*, the primordial indeterminate substrate that is a component of all material being and is conserved in all changes, substantial as well as accidental, in the universe. Aristotle argues for the existence of such a substrate in *Physics* I.7 (189b30–191a22) from an analysis of transmutation or substantial change. See also D3.2.3.

[13] *first mover*: Lat. *motor primum*, the eternal and immovable mover required, according to Aristotle in *Physics* VIII.6 (258b10–260a19), to explain all movement or change in the universe. This and the previous example are invoked by Zabarella and Balduino in their commentaries on the *Posterior Analytics*; see Lat. Ed. 130. See also D3.2.3.

[14] *usually not proved but manifested*: Lat. *non solere probari sed ostendi*, not demonstrated but simply shown to the person who does not see them by the process described in F2.1.4 n. 11; see Sec. 4.6b of *Galileo's Logic of Discovery and Proof*.

[15] *first and immediate principles*: already mentioned in F2.1 and explained at length in D2.2 and D2.3.

[16] *subalternating science*: a science that is higher than another (called the subalternated science) in the sense of being more abstract and more certain. For Aristotle mathematics stands in this relation to physics and so can supply principles for "mixed sciences" such as astronomy, optics, and mechanics, the ancient counterparts of mathematical physics. See Secs. 3.3c and 3.4 of *Galileo's Logic of Discovery and Proof*.

[17] *dialectics*: the branch of logic concerned with probable reasoning, whose canons are set forth by Aristotle in the *Topics*; see Sec. 3.6 of *Galileo's Logic of Discovery and Proof*.

[18] *sought*: Lat. *quaesita*, in the sense that principles are the end of the reasoning process when one argues from effect to cause, whereas they are the beginning of the reasoning process when one argues from cause to effect.

F3: *On Foreknowledges of Subjects*

F3.1: What does Aristotle mean by the "is" of the subject? *For background, see Secs. 3.1–3 and 4.3 of* Galileo's Logic of Discovery and Proof, *for applications, Secs. 6.5b and 6.8c*

¹ {of the Subject}: this expression is missing in Galileo's manuscript; Galileo left a space for it but then failed to return and fill the space in, thus omitting it inadvertently from the title of the disputation.

² *Aristotle*: the reference is to the first chapter of the first book of Aristotle's *Posterior Analytics*, where Aristotle maintains that to have demonstrative knowledge of the unit, one must first know what the term "unit" means and whether or not a unit in this understanding exists (71a16–17). The question thus relates to previous knowledge, i.e., to foreknowledge, as this is required to demonstrate properties of some subject such as a unit.

³ *"is"*: Lat. *esse* in the first occurrence, and *an sit* in the second. Literally *esse* means "to be," and sometimes it is used as a synonym for "being," but it is employed in scholastic Latin in a more technical sense to indicate existence, following the usage of St. Thomas Aquinas. Similarly *an sit* literally means "Is it," "Whether it be," or "Whether it exists," but it is also used as a synonym for existence. Thus the question reads: What does Aristotle mean by existence when he says that the existence of the subject must be foreknown?

⁴ *subject*: the subject of a demonstration whose conclusion is usually expressed in the form S is P, where S stands for the subject and P stands for the predicate or property attributed to it.

⁵ *subject*: subject can be taken for the subject of a demonstration, as above, or it can be taken for the subject of a science. In the latter meaning it is sometimes used interchangeably with object, in the sense that the subject of a science is also the object of its investigations.

⁶ *properties*: predicates in the sense of proper attributes. Thus it is a property of a triangle that the sum of its interior angles is equal to two right angles.

⁷ *metaphysics*: the highest of the speculative sciences, treating of being in the most general sense (*ens commune*), as opposed to quantified being (*ens quantum*), studied in mathematics, and changeable being (*ens mobile*), studied in natural science. See Sec. 3.3b of *Galileo's Logic of Discovery and Proof*.

⁸ *particular sciences*: those with special subject matters, such as astronomy, treating of the heavens, or meteorology, treating of phenomena in the earth's atmosphere.

⁹ *the "is" of existence*: Lat. *esse existentiae*, usually taken as a synonym for existence. It is opposed to *esse essentiae*, usually taken as a synonym for essence.

¹⁰ *actual or potential*: these terms are explained in the examples that follow. Actual existence means existence here and now, whereas potential existence means merely the ability to exist or to come into being. A live human being has actual existence; a fertilized human egg has actual existence as an egg but only potential existence as a human being.

¹¹ *authorities*: the reference here is to authors (*auctores*) who have treated this problem – generally Greek, Arab, or Latin commentators on the *Posterior Analytics*. Many of them are identified by name in what follows, along with references to the *loci* where they discuss Aristotle's text.

¹² *for acquiring a science at the start*: that is, at the time when a science is first being developed by those who initiate it, its discoverers. The problem is slightly different for those

who later acquire a science from those who first discovered it and so may be said to learn it from them. See Sec. 3.1 of *Galileo's Logic of Discovery and Proof*.

[13] *in the development of a science*: Lat. *in progressu scientiae*. In Galileo's terminology, *progressus* as applied to a science has a special technical sense explained in Sec. 4.9c of *Galileo's Logic of Discovery and Proof*. Here he is simply referring to what is required for the growth of any science, whether this comes about in those who initiate it or in those who learn from them (cf. D3.3.10 nn. 18–19).

[14] *second "is"*: i.e., the "is" of existence, as opposed to the "is" of essence.

[15] *certain authors*: these are not identified by Galileo. Vallius-Carbone name them as Marc Antonio Zimara, Girolamo Balduino, Apollinaris, and Giles of Rome.

[16] *the "is" of the essence*: Lat. *esse essentiae*, i.e., the essence of the subject, or its quiddity, usually formulated in its definition. Thus, to demonstrate properties of a triangle one must first know the essence of a triangle, that is, what a triangle is, namely, a three-sided plane figure.

[17] *essence before...existence*: this statement presupposes the Thomistic real distinction between essence and existence in creatures. In Aquinas's teaching existence is an act conferred on a creature by the Creator that complements and perfects its essence by putting it among actual existents. See F2.4.1 n. 8.

[18] *determinate quiddity*: that is, precise definition.

[19] *of others*: these are not identified by Galileo or by Vallius-Carbone, but Vallius mentions Domingo de Soto as one such author in VL2: 164.

[20] *many moderns*: Lat. *multorum recentiorum*, possibly a reference to authors still living and on this account not mentioned by name.

[21] *actual "is" of the existence*: Lat. *esse existentiae actuale*, i.e., existence in the real world, not merely in the mind, and not potentially but in full actuality.

[22] *at all times*: Lat. *semper*. The word *semper* occurs twice in this sentence; the first occurrence was apparently missed by Galileo in adapting his text from the source from which he worked, since it is inserted in the text interlinearly. This question has significant implications for the development of Galileo's science of motion, particularly for his treatment of uniformly accelerated motion. The sense of the query is this: do the things of which we seek scientific knowledge have to be real and existent at the very time our scientific knowledge exists? For example: can there be a science of roses in winter, when all roses are dead; can there be a science of snow in mid-summer, when it is too hot for snow actually to exist? See D2.1.9.

[23] *the "is" of the {essence}*: Lat. *esse essentiae*; Galileo wrote *esse existentiae* here, obviously a copying error and so corrected for sense.

[24] *the "is" of the* {essence}: see the previous comment.

[25] *there cannot be any demonstration*: usually the middle term of a demonstration states the definition of the subject; if the definition of the subject, i.e., the "is" of its essence, is not known, no middle term will be available and thus there cannot be a demonstration. See n. 16 above.

[26] *in real sciences*: Lat. *in scientiis realibus*, i.e., in sciences concerned with real, extramental beings, such as those in the world of nature. Real sciences are here implicitly differentiated from rational sciences, those concerned with beings of reason, which have existence in the mind only and not in the world of nature. Logical entities are commonly regarded as beings of reason. See Sec. 2.3 of *Galileo's Logic of Discovery and Proof*; also n. 34 below.

[27] *at least. . .impediments*: this expression was omitted by Galileo in his first draft of the passage, and then inserted at its proper place in the margin of the manuscript. For the importance of the marginal insert as evidence of the derivative nature of Galileo's composition, see *Galileo and His Sources*, 41–42. The point of the qualification, as explained by Carbone in his version of the passage (*Additamenta*, 46vb–47ra), is that one need not know of the actual existence of the subject of demonstration for all times and places and under all conditions whatever. It is sufficient to know, for example, that roses actually exist in the summer in the earth's northern hemisphere, provided that there is no blight in the region that would kill them; under these suppositions it is possible to have a science of roses, even though no roses may actually be existent here and now. Galileo was interested throughout his life in *impedimenta*, i.e., accidental causes that interfere with the phenomena of nature, and devoted much of his experimental activity to eliminating them. A good part of his study of naturally accelerated motion, described on the Third Day of his *Two New Sciences*, was in fact directed at identifying *impedimenta* such as friction and air resistance that would cause the actual fall of bodies to deviate from the uniform acceleration imparted to heavy bodies by nature. On the supposition of such impediments being removed, one could have a true science of naturally accelerated motion and demonstrate properties of it as a subject. See Sec. 4.2c of *Galileo's Logic of Discovery and Proof*.

[28] *human sciences*: i.e., sciences as possessed by human beings, and thus to be distinguished from those known only to angels or to God. See Secs. 2.7a and 3.1 of *Galileo's Logic of Discovery and Proof*; also n. 45 below.

[29] *an ["is"] of actual existence*: Lat. *esse actualis existentiae*; in writing this Galileo omitted the *esse*, here inserted for sense.

[30] *the first of all [scientific questions]*: according to Aristotle in the beginning of Book 2 of the *Posterior Analytics* (89b23–25), there are only four scientific questions that can be asked about any subject. These were usually given in Latin by the scholastics as follows: *an sit* (Is it?); *quid sit* (What is it?); *qualis sit* (What kind is it?); and *propter quid* (Why is it of this kind?). Thus the question *an sit* is the first of all scientific questions.

[31] *This is apparent. . .eternity of motion*: see F2.34.3 and D3.2.3. Here are enumerated exceptions to the general principle established in the second conclusion, paragraph [9] above. Although the total subject of the science must be foreknown to exist actually in a general way, this requirement does not entail that everything included under the total subject (spoken of in the following question as a "partial subject") need be foreknown in detail. For example, to have a science of motion or change, one would have to know the principles that are required for motion or change in general, without having to be aware that a special type of change, known as "substantial change," requires protomatter or *materia prima* as its proper subject. Similarly, one might be aware that all change requires an agent, without knowing in detail that a special agent, or "prime mover," might be necessary to explain the eternal movement of the heavenly bodies. Or again: one would need to foreknow that local movement, or change of place, actually exists in nature in order to have a science of motion, but one would not have to know that there is such a thing as naturally accelerated motion at the outset of the science; indeed, one might prove the existence of the latter type of motion, under certain conditions, as the science later develops. Much of Galileo's discussion throughout the Third and Fourth Days of the *Two New Sciences* seems directed at elucidating this point.

[32] *of demonstration in its proper sense*: i.e., demonstration of the reasoned fact, as opposed

to demonstration of the fact, the type discussed in the previous note. For fuller details see Secs. 4.4b and 4.9a of *Galileo's Logic of Discovery and Proof*.

[33] *rational sciences, if there be such*: in D2.1.8 Galileo gives the reason for the qualification "if there be such," namely, that science must be concerned with real beings that are true in an ontological sense. On this account Vallius-Carbone (and Vallius himself), while occasionally referring to logic as a science, prefer to characterize it simply as an instrumental habit; see Sec. 2.4a of *Galileo's Logic of Discovery and Proof*.

[34] *its "is" as an object*: Lat. *esse obiectivum*, i.e., objective existence, or existence solely as an object of consideration by a knowing power such as the intellect, and not existence in a real or extramental sense.

[35] *sciences abstract from existence*: an objection based on the abstractive theory of the sciences deriving from Boethius and Aquinas. In this view the concepts on which a science is based are abstracted from individual, sensible existents, and in the process are universalized; thus they consider the essences or universal features of things, not their singular or particular instantiations, and in this sense do not consider the existence of this or that individual. For an explanation of the knowledge process on which this theory is based, see Sec. 2.3 of *Galileo's Logic of Discovery and Proof*.

[36] *sciences are not concerned with contingent matters*: this is a variation of the objection in paragraph [12], and the reply is the same except that essences or universal features of things are replaced by species, which are not contingent in the way in which individual existents are. On the difference between necessary and contingent matters, see Sec. 3.5a of *Galileo's Logic of Discovery and Proof*.

[37] *on the supposition of there being a universe*: Lat. *supposito universo*, that is, if there is to be a universe made up of many species at any particular time, then those species are necessary under that supposition. This is an instance of suppositional reasoning, frequently employed by Galileo; see Sec. 4.2 of *Galileo's Logic of Discovery and Proof*.

[38] *science does not consider individuals*: yet another variation of the objection given in paragraphs [12] and [13], and the reply is again the same, except that nature now becomes the universalizing principle. Medieval writers such as Jean Buridan regarded an argument of this type as one made *ex suppositione naturae*, i.e., on the supposition of nature – an observation which ties this reply to that of the previous argument.

[39] [...]: the ellipsis indicated here is a blank space of about half a line that occurs in the manuscript at this point. Galileo apparently had difficulty abbreviating the argument in his source and so left a space to be filled in later. His lacuna here occupies six lines in Vallius-Carbone, which translate as follows: "Since therefore species are conserved in individuals, there must be some individual in which they exist. Thus the universal nature is foreknown, not in the abstract but in some singular individual, though not one determinately assigned, and that is in no way repugnant to the sciences" [Lat. Ed. 140].

[40] *removing the impediment of the divine will*: note here that God's efficacious will, which could annihilate the universe at any time, is regarded as an *impedimentum* to the work of the natural philosopher; thus even this impediment has to be removed by an appropriate supposition if one is to reason as a natural scientist.

[41] *angels*: the objection is raised to differentiate human sciences, paragraph [9] above, from those that may be acquired by spiritual creatures. Human beings obtain their knowledge of natural kinds through a process of abstraction from individual existents perceived in sense experience, and thus actual existents are necessary for their acquisition of a science, as noted

in paragraph [3]. Angels, on the other hand, obtain knowledge through infused species and without sense impressions; therefore they are not limited as are humans in this regard. Again see Secs. 2.7a and 3.1 of *Galileo's Logic of Discovery and Proof* for fuller background.

[42] *human sciences*: i.e., sciences acquired by men, not those possible for angels; see nn. 28 and 45.

[43] *properties can be present*: the objection makes the point that potential presence, as opposed to actual presence, is sufficient for knowledge of the property, and therefore potential existence might also be sufficient for knowledge of the subject – a restatement of the argument proposed as the second opinion in paragraph [5] above. The reply, rather than rejecting this opinion outright, acknowledges some element of truth in it.

[44] *mathematics...goodness*: this Aristotelian maxim was quoted frequently by Galileo in his various writings. Some authors used it in a pejorative sense to calumniate mathematicians and their work – examples would be Alessandro Piccolomini and Benedictus Pererius, the latter a Jesuit who taught at the Collegio Romano before Vallius. Following Vallius's notes, Galileo employs the maxim but turns it to good use to show that mathematics is a human science concerned with existence, as explained in the following note.

[45] *human science*: that is, a science derived from sense experience. It is also a real science, and not merely a rational science like logic concerned with an *esse obiectivum*, as indicated in paragraph [11], n. 34. This being the case, it presupposes the actual extramental existence of quantified beings in order to establish its subject, even though it abstracts from such existence when demonstrating their properties.

F3.2: Can a science demonstrate the existence of its adequate subject?
For background, see Secs. 3.3, 4.3, and 4.4 of Galileo's Logic of Discovery and Proof, *for applications, Secs. 6.5 and 6.8*

[1] *adequate subject*: Lat. *obiectum adaequatum*, an instance where subject and object are taken to be equivalent; see F3.1.1 n. 5. Many different kinds of subject/object were mentioned and distinguished in Jesuit commentaries on Aristotle's works, viz., principal subject, adequate subject, total subject, partial subject [see Sec. 4.3a of *Galileo's Logic of Discovery and Proof*]. The adequate subject of a science is a subject that is equivalent to, or may be equated to, all of the objects that are considered in a science; thus nature can be said to be the adequate subject of physics or natural science.

[2] *total subject*: Lat. *subiectum totale*, one roughly equivalent to the adequate subject of the science but stressing the proper formality under which the science considers it. In the case of physics, this was taken by the Jesuits to be the natural body, which includes all bodies studied in natural science [Sec. 4.3a of *Galileo's Logic of Discovery and Proof*].

[3] *total object*: Lat. *obiectum totale*, used interchangeably with total subject.

[4] *partial science*: Lat. *scientia partialis*, the portion of a science that considers a partial subject; for example, if the atmosphere is only a part of the sublunary region, the science that treats this part of terrestrial nature, meteorology, is a partial science with respect to the science that treats all of the elementary bodies (for Aristotle, the *De caelo et mundo*), which therefore can be regarded as a total science. Partial and total are thus correlatives. The science of the elements is a partial science with respect to the total science of physics, and the science of earth (as one of the four Aristotelian elements) is a partial science with respect to

the total science of the elements. See the following two notes.

⁵ *total science*: Lat. *scientia totalis* – see the previous note. In a more proper sense, when discussing the specification of the speculative sciences, Vallius-Carbone argue that here are only three such total sciences, namely, physics, mathematics, and metaphysics. The basis of this division is explained in Sec. 3.3b of *Galileo's Logic of Discovery and Proof*.

⁶ *a partial science...of its subject*: a partial science has a partial subject or object, and the possibility exists that it can demonstrate the existence of its partial subject. For example, although earth and water and air give evidence of their extramental existence, it is not completely clear that fire exists as an element, and thus it would be necessary to prove the existence of fire if one were to have a science of this element (cf. D3.2.3). This question apparently had important ramifications for Galileo in the later development of his science. For example, the second of the "two new sciences" he proposed to develop in his masterwork of 1638 was the science of local motion; its total subject would therefore be this type of motion, motion according to place (*motus localis*), and as he proposed it, it would be composed of three partial subjects that could count as its species: uniform motion, naturally accelerated motion, and a combination of these two, projectile motion. For each of these partial subjects, in turn, it would be necessary to offer experimental or *a posteriori* proofs of their extramental existence.

⁷ *the [question] "What is it" of its subject*: Lat. *quid sit sui subiecti*, the definition or the quiddity of the subject; see F3.1.4 n. 16 and F3.1.8 n. 25. In the second book of the *Posterior Analytics* Aristotle argues that no science can demonstrate the definition of its subject, but that nonetheless it can manifest that definition in a demonstrative way by showing the relationships that obtain among its causal components; see chapter 10, 93b37–94a9.

⁸ *manifested*: Lat. *ostendi*; see F2.1.4 n. 11, F2.4.4 n. 14.

⁹ *in a higher science*: Lat. *in [scientia] superiori*, that is, in a subalternating science as explained in F2.4.5 n. 16. For fuller details, again see Secs. 3.3c and 3.4 of *Galileo's Logic of Discovery and Proof*.

¹⁰ *are correlatives*: Lat. *sunt relativa*, basically the same argument as applied to cause and effect, F2.3.1; see also D3.3.7–8.

¹¹ *of the principal subject*: Lat. *subiecti principalis*, the preeminent part of the total subject; in his notes on Aristotle's *De caelo et mundo* Galileo identifies the heavens (*caelum*) as the principal subject (*obiectum principalitatis*) of that treatise, GG1: 16. See Sec. 4.3a of *Galileo's Logic of Discovery and Proof*.

¹² *here*: Galileo failed to identify the particular text of Aristotle on which the argument in part [c] of paragraph [1] is based. A clue is given in Vallius-Carbone, however, who point to the sixth book of the *Metaphysics*, ch. 1 (1025b17–18), as the source of the difficulty [CA49ra; Lat. Ed. 146].

¹³ *quiddity*: Lat. *quid sit*, essence or definition, here again opposed to *an sit*, existence; see F3.1.4 nn. 16–18.

¹⁴ *this text*: i.e., that identified in n. 12 above.

¹⁵ *habitus*: habit in the sense of second nature; a scientific habitus perfects the intellect, enabling it to function in an effortless way with its subject matter in much the same way as the virtue of justice perfects the will of a just man and enables him similarly to act justly in his dealings with others. This notion lies behind Vallius-Carbone's distinction between actual science and habitual science, as explained in Sec. 3.2 of *Galileo's Logic of Discovery and Proof*.

[16] *a higher science*: a reference to the comparison and subalternation of the speculative sciences, explained in Sec. 3.3c of *Galileo's Logic of Discovery and Proof*.

F3.4: Can a science demonstrate the existence of its partial subject? *For background, see Secs. 2.5 and 4.3 of* Galileo's Logic of Discovery and Proof, *for applications, Secs. 6.6b, 6.7b, and 6.8*

[1] Galileo enumerates this as the "Fourth Question" but actually it is the third of the disputation. From a study of his handwriting and the contents of F3.2 and F3.4 one can establish with near certainty that he did not skip a question here but merely made a mistake in numbering the questions; for details, see Lat. Ed. 146 (16.17–20), 147 (17.1–2). Since much the same terminology is employed here as in F3.2, the reader should consult the notes for that question to supplement those given below.

[2] *Some*: Lat. *Aliqui*, not identified by Galileo. Vallius-Carbone note that their objections derive from Averroes, who held the opposite opinion but listed these and other arguments in order to refute them.

[3] *Others*: Lat. *Alii*, again not named by Galileo but identified by Vallius-Carbone as Averroes, Grosseteste, and Zimara. Other proponents are specified by Lorinus; see Lat. Ed. 148–149.

[4] *habitus of a higher science*: see F3.2.10 nn.15–16.

[5] *to pass from one genus to another*: Lat. *de genere in genus transcendere*. The prohibition here is usually referred to as that against *metabasis* or "passing over" from one genus to another, voiced by Aristotle in A.7 of the *Posterior Analytics* (75a37–39), where he states that one cannot use arithmetic to demonstrate a theorem in geometry. In this particular case arithmetic, whose subject is discrete quantity, is regarded as being concerned with a higher genus than geometry, whose subject is continuous quantity. Following this expression there is an addition of six words in Galileo's composition; these are omitted here as representing an incomplete thought and so not translatable – for details, see Lat. Ed. 149.

[6] *the actuality of the subject*: Lat. *actus ipsius subiecti*, an application of St. Thomas's teaching on potency and act as the basic components of all being. In this view essence is related to existence as potency is to act; thus existence is the ultimate actuality of the subject, presupposing its quiddity as the potency it actuates. See F2.4.1 n. 8, F3.1.4 n. 17.

[7] *[pointless]*: Lat. *vana*, omitted inadvertently by Galileo in this sentence but appearing correctly in the following sentence.

[8] *to investigate the parts of its definition*: explaining how this investigation is to be carried out takes up a large part of the second book of the *Posterior Analytics*. A definition may be viewed as composed of two parts, the genus and the differentia, and this would be known among scholastics as a metaphysical definition, or alternatively it may be viewed as composed of four parts, Aristotle's four causes (formal, material, efficient, and final), in which case it would be called a physical definition.

[9] *first text*: that is, *Metaphysics* E.1, 1025b3–18.

[10] *all causes…the formal cause*: a further explication of the parts of a definition as noted in n. 8 above. Although all four causes pertain to the definition or quiddity in a broad sense, the formal cause is preeminent among them and thus serves best to characterize the quiddity. The statement is corroborated in D2.2.4, where Galileo points out that demonstrations of

the reasoned fact are more perfect the more they proceed from formal causes, which are more intrinsic to the thing.

F3.5: Can a science manifest the real definition of its subject? *For background, see Secs. 2.5, 4.3, and 4.4 of* Galileo's Logic of Discovery and Proof, *for an application, Sec. 6.1*

¹ *real definition*: Lat. *quid rei*, definition of the thing, as opposed to the *quid nominis* or definition of the term, also called the nominal definition; see F2.2 n. 1, F3.1.4 n. 18, F3.2.1 n. 7.

² *explain...apodictically*: Lat. *reddere propter quid*, that is, give a demonstration *propter quid* of the subject's existence. Usually "demonstration *propter quid*" is translated "demonstration of the reasoned fact," as opposed to "demonstration *quia*," which is translated "demonstration of the fact." The fact of the subject's existence is not at question here; the problem is the "why" of its existence and how the real definition can supply this.

³ {*existence*}: In writing the title of his question Galileo unaccountably left out the word for existence; that it was intended is clear from the first sentence of paragraph [3] in his text.

⁴ *a posteriori only*: Galileo's first conclusion in this paragraph, that a science can show real definitions of its partial subjects, is unproblematic in view of his conclusions in F3.4; his second conclusion relating to the total subject is more controversial in view of the related discussion in F3.2. He thus qualifies the second conclusion by stating that the demonstration manifesting the real definition of the total subject can only be *a posteriori*. Vallius-Carbone explicate this by adding how this can be done: "through some effects that are more known." An example might be the definition of nature given by Aristotle in *Physics* II.1, 192b21–23, based on its effects as described in 192b9–21.

⁵ *as we have explained above*: i.e., in F3.4.8.

⁶ *existence in common*: Lat. *existentia in communi*, the way in which existence is considered by the metaphysician, explained in F3.4.8.

⁷ *one must beg the question*: Lat. *petere principium necesse est*. Note here that this is the first mention of a *petitio principii* in Galileo's appropriated questions – important because it was probably a concern over begging the question in his *Theorems on the Center of Gravity of Solids* that prompted his interest in Vallius's lecture notes. Other occurrences of the expression are at D2.3.5, D3.1.5–6, D3.1.15, and D3.3.2.

⁸ *in the minor premise*: i.e., in the second premise stating that every man is a rational animal. Precisely how this begs the question is not clear from Galileo's exposition, since he will give exactly the same syllogism, with the same minor premise, in paragraph [5]. Vallius-Carbone make the point more explicitly: the minor premise as here stated is understood to mean that every man exists as a rational animal, taking the "is" of the premise to mean "exists"; therefore it begs what was to be proved in the conclusion, namely, that every man exists. See n. 11 below.

⁹ *apodictic proof*: that is, proof in the sense of that formulated in a most powerful demonstration. How most powerful demonstration can be said to provide *propter quid* proof of existence is explained at length in D3.1.

¹⁰ *of which it is the actuality*: Lat. *cuius est actus*; see F3.4.4 n. 5.

¹¹ *the word "is"...meaning of existence*: this second formulation of the syllogism, though

expressed with the same premises as that in paragraph [3], is meant only to explicate the content of the preceding enthymeme and not to make any existence claim in the minor premise. Thus the "is" is to be understood only as a logical copula, not as an ontological affirmation.

[12] *To the objections*: a superfluous addition made by Galileo and not found in his source. The difficulty raised in paragraph [2] is answered in the same paragraph, that in paragraph [3] in paragraph [5].

F3.6: What does Aristotle mean by foreknowledge of the subject's quiddity? *For background, see Secs. 4.1, 4.3, and 4.7 of* Galileo's Logic of Discovery and Proof, *for an application, Sec. 6.7b*

[1] *What does Aristotle mean*: that is, at the beginning of the *Posterior Analytics*, 71a12–14, in a passage included in text 2 according to Galileo's enumeration of text numbers; see F2.2.1 n. 2.

[2] *quiddity that is said [of the subject]*: Lat. *quid est quod dicitur*. The Latin of the version of James the Venetian Greek is a cryptic but faithful rendering of the Greek text. It reads as follows: Dupliciter autem necessarium est praecognoscere; alia namque quia sunt, prius necesse est opinari, alia vero, quid est quod dicitur intelligere oportet... Difficult to translate, this passage was generally taken to mean that two things must be known about the subjects of demonstration, "that" they are (*quia sunt*) and "what" it is that is said of them (*quid est quod dicitur*). Some commentators took the quid of the latter expression to refer to the *quid rei* or real definition, others to the *quid nominis* or nominal definition; see F2.2 n. 1, F3.2.1 n. 7.

[3] *Cajetan*: the only recent expositor mentioned by Galileo; Vallius-Carbone identify Soto and Zabarella as others who held this position.

[4] *many moderns*: to Themistius and Philoponus Vallius-Carbone add Giles of Rome, Paul of Venice, and Apollinaris; Lorinus attributes the position to contemporary Averroists, *recentiores Averroistae*.

[5] *in some demonstrations*: Galileo here cites demonstrations *a posteriori* as exceptions that do not require foreknowledge of real definitions; in his *Logica* Vallius points to mathematical demonstrations as those wherein foreknowledge of real definitions "is not required in any way" [VL2: 141].

[6] *quiddity presupposes existence*: a familiar refrain in this treatise; see F2.4.1 n. 8, F3.1.4 n. 17, F3.4.4 n. 5.

[7] *"that which the thing was to be"*: Lat. *quid quod erat esse rei*, a literal translation of Aristotle's *to ti en einai* (*Metaphysics* 988a34, 1013a29, 1037a22) attributable to William of Moerbeke, though with the first two words interchanged. The expression was much used by St. Thomas, usually to signify the formal cause "through which is known what the substance of a thing is" (*In I Meta.*, lect. 11; see also *In V Meta.*, lect 2; *In VII Meta.*, lect. 11).

[8] *"the quiddity that was"*: Lat. *quid quod erat*, an abbreviation of the fuller expression found in the previous note.

[9] *[and]*: omitted by Galileo and required for sense.

[10] *"directing."..."acting"*: see the comment above at F2.2.5 n. 8 and the fuller discussion in Sec. 4.1 of *Galileo's Logic of Discovery and Proof*.

[11] *to experts*: Lat. *sapientibus*, an expression based on the scholastic distinction between principles evident to all (*per se nota omnibus*) and those evident only to experts (*per se nota sapientibus*), that is, those who have a competent grasp of a particular subject matter.

[12] *"How many things should be foreknown"*: Lat. *Quot sint praecognita*, the title of a question pertaining to the first disputation on foreknowledge, missing from Galileo's treatment. Galileo's entire treatise, as its title indicates, is concerned not only with foreknowledges (*praecognitiones*) but also with things that must be foreknown (*praecognita*). The *praecognita* would seem to be three in number, corresponding to the objects of consideration in the three disputations into which the treatise is divided, namely, principles, subjects, and properties. Precisely as foreknown, they alternatively may be viewed as *praecognitiones*, as Galileo implies in the titles of his three disputations. Following Aristotle's usage, however, the *praecognitiones* may be differentiated from the *praecognita*, since in text 2 Aristotle seems to state that the *praecognitiones* are only two in number, whereas the *praecognita* are the three just mentioned. The two foreknowledges to which Aristotle refers there can be gathered from the passage given in Latin in n.2 above and introduced by the expression (*Dupliciter...praecognoscere*); these are the *quia sunt*, concerned with questions of existence, and the *quid est quod dicitur*, concerned with questions of meaning. When one understands this, Galileo observes, one will be able to answer the objections that might be brought against the conclusions of this question. For a fuller treatment, see Sec. 4.1a of *Galileo's Logic of Discovery and Proof*, which discusses kinds of foreknowledges, and Sec. 4.1b, which discusses kinds of foreknowns.

F4: *On Foreknowledges of Properties and Conclusions*

F4.1: Must the existence of a property be foreknown? *For background, see Secs. 4.1, 4.3, and 4.4 of* Galileo's Logic of Discovery and Proof, *for an application, Sec. 6.4b*

[1] *of the property*: Lat. *passionis*, i.e., of the attribute that is predicated of the subject in the conclusion, called on this account a proper attribute or simply a property.

[2] *of the conclusion*: strictly speaking the foreknowledge of the conclusion is already contained in the foreknowledges of the subject and of the property that is predicated of it, and thus it is redundant to treat it as a separate type of foreknowledge. As Galileo indicates in his explanatory remark, however, he is doing so as a matter of convenience. Actually it enables him to make a few notations about the time-sequence involved between knowing the premises and knowing the conclusion in a syllogistic argument, as detailed in F4.2. Indirectly, therefore, a type of "fore-knowing" is involved that can be discussed in a treatise on foreknowledge.

[3] *the existence of a property*: Lat. *An de passione praecognoscendum sit quia est*. The *quia est* here is the equivalent of *an sit* or *quod est*, the first of the scientific questions (F3.1.9 n. 30). By its very nature a property is a type of accidental being and as such its mode of existence is that of existing in another as in a subject. This mode of existence poses special problems for its foreknowledge that do not arise, for example, when discussing the foreknowledge of subjects, most of which are substances that exist by themselves and so do not depend on others for their existence.

[4] *the apodictic reason why it is present*: Lat. *propter quid insit*, the fourth of the scientific questions, F3.1.9 n. 30.

[5] *of a proper attribute*: Lat. *de propria passione*, as in n. 1 above.

[6] *demonstration of the fact*: Lat. *demonstratio quia*, as noted in F3.5 n. 2; for a full treatment of the kinds of demonstration, see the last disputation in the treatise on demonstration, D3; also Secs. 4.4a and 4.9a of *Galileo's Logic of Discovery and Proof*. Vallius-Carbone note that the threefold distinction given here is attributable to Averroes.

[7] *demonstration of the reasoned fact*: Lat. *demonstratio propter quid*.

[8] *most powerful demonstration*: Lat. *demonstratio potissima*.

[9] *convertible with its subject*: Lat. *reciprocam cum suo subiecto*, usually understood to be a property in the strict sense, found solely in the subject and in every instance, as the croak of a frog, the bark of a dog.

[10] *not convertible*: Lat. *non convertibilem*, as round said of the earth.

[11] *nominal definition*: Lat. *quid nominis*, F2.2 n. 1.

[12] *essence*: Lat. *essentia*. The essence of a property would be its *quid rei* or real definition, and a real definition could not be shown of something whose existence was in doubt.

[13] *the property*: Lat. *passio*; in their corresponding conclusion Vallius-Carbone give *passio convertibilis cum subiecto*. That the intended reference here is to a convertible property is clarified by Galileo in paragraph [8].

[14] *not convertible...convertible*: Lat. *non reciproca...reciproca*, as in n. 9 above.

[15] *absolutely and simply speaking*: Lat. *absolute et simpliciter*. The sense is that it is possible for a property such as roundness to be known from experience with nature, say, in an orange, before one comes to know that it is also a property of the earth.

[16] *never...not*: the double negative obscures the sense. What Galileo means is that in a demonstration of the reasoned fact, as opposed to a most powerful demonstration, the property's existence must always be foreknown, since the point of the demonstration is to manifest why that property is found in the particular subject.

[17] *Aristotle makes no mention*: that is, in text 2, as explained in F3.6 n. 1.

[18] *two general kinds of foreknowledge*: see the comment at F.3.6.4 n. 12.

[19] *"acting" knowledge..."directing" knowledge*: see F2.2.5 n. 8.

[20] *real definition*: Lat. *quid rei*. Galileo uses this query to add a corollary about foreknowledge of the real definition of the property, as opposed to its nominal definition, to which a special question was devoted by some commentators. See Lat. Ed. 165.

F4.2: Is the conclusion known simultaneously with the premises? *For background, see Sec. 4.3 of* Galileo's Logic of Discovery and Proof, *for applications, Secs. 5.6a and 6.6b*

[1] *at the same time*: Lat. *simul tempore*, simultaneously.

[2] *with the same priority*: Lat. *[simul] natura*, an additional qualification, since another sense of priority is discernible in events that take place at the same time, as explained in the first notation of paragraph [1]; for example, even though the hand and the pen move simultaneously, the motion of the hand precedes that of the pen, since it is the hand that moves the pen and not vice versa.

[3] *major and minor premises*: in a demonstrative syllogism, usually written M is P; S is M;

therefore S is P, where S and P are the subject and predicate of the conclusion and M is the middle term. The first premise, M is P, is called the major premise because it contains P, a term of broader extension: the second premise, S is M, on the other hand, is called the minor premise because it contains S, a term of narrower extension.

⁴ *as cause to effect*: that is, the way in which the motion of the hand precedes that of the pen, as in the example given in F4.2 n. 2, even though both are temporally simultaneous.

⁵ {*the path*}: Lat. *via*, written incorrectly by Galileo as *quia* and here emended, following the reading in Vallius-Carbone.

⁶ *as things are related...in being...so...in knowing*: Lat. *sicut se habet ad esse, ita ad cognosci*, a principle invoked repeatedly by Galileo; see F3.1.4, D1.1.11, D2.2.7, D2.2.9, D2.5.12, and D3.1.11.

⁷ *by the same act*: The basis for this teaching, as noted by Vallius-Carbone, may be found in Aristotle, but it was extensively developed by St. Thomas and his school, notably Capreolus, Ferrariensis, and Cajetan; see Lat. Ed. 168–169.

⁸ *I say, third...*: This conclusion is essentially the same as the third in Vallius-Carbone, except that they state it as probable rather than certain.

⁹ {*assent*}: Lat. *assensum*, emending Galileo's *sensum*, which is unintelligible in this context and probably an error in copying.

¹⁰ *its premises* {*affirmative*} *and the conclusion* {*negative*}: correcting Galileo's text, which reads "its premises negative and its conclusion affirmative," clearly a copying error, since the phrase does not cohere with the rest of the sentence. Again, one can have a valid syllogism with a negative conclusion and an affirmative premise, but not one with a negative premise and a positive conclusion, for this would violate the rules of the syllogism.

¹¹ {*then*}: Lat. *tunc*, emending Galileo's *tum*, obviously a writing error.

¹² {*at the same time*}: Lat. *simul*, emending Galileo's *simus*, another slip of the pen.

¹³ *every cause that is sufficient to produce its effect...*: a statement that is important for understanding Galileo's conception of causality, as explained in Sec. 4.5 of *Galileo's Logic of Discovery and Proof*. A similar principle is employed in D2.6.6; see also D2.6.6 n. 17.

¹⁴ {*not*}: supplying the negation, missing from Galileo's text, but required for sense.

GALILEO GALILEI

TREATISE ON DEMONSTRATION

MS 27, Fols. 13r – 31r

Translated, with Notes, by William A. Wallace
From the Latin Edition of W. F. Edwards and W. A. Wallace

[TREATISE ON DEMONSTRATION][1]

[This is] the treatise on demonstration, having omitted the definition of science, which Aristotle treated quite judiciously [...],[2] beginning his treatise with it, so that knowing the end of demonstration, which is science, he might better and more perfectly cast light on its nature and properties, about which we will have much to say in the following treatise.[3] We proceed to treat demonstration itself, which contains three disputations within it: the first, on the nature and importance of demonstration; the second, on its properties; the third, on its species. When these are done and completed, nothing that can be known about demonstration will remain to be desired.

[D1] *First Disputation*[4]: *[On the Nature and Importance of Demonstration]*

[D1.1] *First Question: On the definition of demonstration*

[1] Note, first: there are many species of demonstration,[5] as is apparent from Aristotle in this book, namely: ostensive; to the impossible; of the fact; of the reasoned fact; and most powerful. An ostensive demonstration[6] is one that proves something true from true principles. A demonstration to the impossible is one that leads from the concession of one impossibility to another that is more known. A demonstration of the fact is one that proves something from an effect or from a remote cause. A demonstration of the reasoned fact is one that demonstrates a predicate of a subject through true and proper[7] principles. A most powerful demonstration is one that manifests some first and universal property of an adequate subject through proper and proximate[8] principles; if any of the foregoing conditions be absent, it will not be most powerful. Here I make no mention of the division into universal and particular, affirmative and negative.[9] The problem at present for us concerns demonstration of the reasoned fact and especially most powerful demonstration.

[2] Note, second: demonstration can be taken in two ways, either as it is a type of illative discourse[10] or as it is an instrument of scientific

knowing.[11] Taken in the first way it has four causes: the efficient cause, the intellect; the material cause, [which is twofold], the "in which," the possible intellect[12] as receptive of intellection within itself, the "from which," terms and propositions; the formal cause, proper arrangement according to mode and figure; and the final cause, [again twofold,] primarily, actual science of the conclusion, secondarily, habitual science[13] of it. Taken in the second way, the efficient cause, again the intellect; the material cause, the subject, the predicate, and the middle term; the formal cause, the necessary relationship of the middle term to the subject, the predicate, and the conclusion; and the final cause, the same relationship.

[3] Note, third: two definitions of demonstration can be gathered from Aristotle in this book: one is that it is a syllogism producing science[14]; the other that it is a syllogism consisting of premises that are true, first, immediate, more known than, prior to, and causes of the conclusion.

[4] With regard to the first definition, note, first: I put "syllogism" in place of the genus, because demonstration has this in common with the probable and the sophistic syllogism; I say "in place of the genus" because "syllogism" is not properly the genus, since it is composed of material species and not of formal species.[15]

[5] Note, second, that where I put "producing science," this is to be understood either instrumentally, because the proper cause of science is the intellect, or dispositively,[16] because demonstration is a kind of condition that is necessary for us to know scientifically.

[6] Note, third, that where I put "science," this is to be taken for perfect science,[17] since a perfect science is one that yields knowledge through causes.

[7] A first objection against this definition: let there be someone who might have a particular perfect demonstration in his mind but would not assent to its conclusion; demonstration in such a person would not produce science and nonetheless it would be a true demonstration, as we have supposed; therefore [a demonstration is not a syllogism producing science]. Some reply: it is of the essence of demonstration only that it be apt to generate science, not that it actually produce it. Others reply, and more to the point: to have a true and perfect demonstration there must be certain and evident assent to the conclusion, and without this one would not really have a demonstration.

[8] A second objection: the effect of demonstration is to know scientifically; therefore this definition is not based on the final cause. Confirmation: because effect and cause are opposed to each other; but to

know scientifically is the effect of demonstration; therefore it cannot be its final cause. I reply, with Aristotle in the *Metaphysics*: nothing prevents the same thing from being both a final cause and an effect, though in different ways. This is obvious in the case of health, which is both an effect of drinking medicine and its final cause; it is also true in the case of science, which is an effect from the viewpoint of being produced by demonstration and a final cause from the viewpoint of demonstration being on its account.

[9] A third objection: if this were so, it would follow that knowledge of the conclusion would be superior[18] to knowledge of the premises, for the end is superior to things ordered to it; but knowledge of the conclusion is the end of knowledge of the premises; therefore [knowledge of the conclusion is superior to that of the premises]. I reply: viewed in this way the argument is valid; absolutely and simply understood, however, knowledge of the premises is superior because it is the efficient cause of knowledge of the conclusion.

[10] A fourth objection: demonstration is defined by Aristotle through scientific knowing, and scientific knowing through demonstration; therefore, since a thing, precisely as itself, cannot be defined through itself, the definition is invalid. I reply: a thing, precisely as itself, cannot be defined through itself considered as itself, but it can be defined by itself considered in a different way. And in this way Aristotle defines science through demonstration as an effect through its efficient cause, and he defines demonstration through science [as an end] through its final cause.

[11] With regard to the second definition,[19] note that I say "a syllogism consisting of true premises,"[20] because truth can only be inferred from truth; I say "of first and immediate,"[21] understood either actually or virtually; I say "of more known,"[22] understood both with respect to nature and with respect to us – for premises that are causes of things cannot be taken to prove anything unless they are more known than it is; and I say "from antecedents and causes of the conclusion,"[23] for things are related in knowing as they are related in being; whence [as causes are prior to the conclusion in being, they must also be prior to it in knowing].

[D1.2] *Second Question: Is demonstration the best[1] of all instruments of scientific knowing,[2] or is definition?*

[1] Note, first: all instruments have been invented so that some end might be attained through them, and for this reason their number should be

taken from their different ends. From this it follows that there will be as many instruments of logic as are necessary for the perfect direction of the operations of the intellect.

[2] Note, second: of the instruments that serve knowledge some are natural, others adventitious.[3] The natural instruments are the intellect, memory, intellectual light,[4] and sense. Among adventitious instruments, some serve knowledge mediately, such as division and method; others immediately[5] but imperfectly, such as the probable syllogism, induction, enthymeme, etc.; others immediately and perfectly, such as demonstration and definition.

[3] Note, third: if we focus on instruments that serve to direct the perfect operation of the intellect in some way, all are agreed that these are six in number: definition; demonstration; division; proposition[6]; argumentation in general, containing under it the probable syllogism, induction, enthymeme, etc.; and method.[7] Proof of this: these six instruments are so related that one cannot be contained easily within the other, and all others are reducible to them; therefore [six instruments are necessary and sufficient]. Proof of the minor: resolution and composition[8] are reducible to demonstration; and demonstration, since it is a perfect instrument, cannot be contained under argumentation in general. If, on the other hand, we focus on instruments that immediately serve operations of the intellect but of a more perfect kind, these are two, argumentation and definition. Proof of this: we know substance through definition, accidents through various kinds of argumentation; also the four questions that are enumerated by Aristotle at the beginning of the second book of the *Posterior Analytics* can be answered by two instruments, for the "What is it?" can be known by definition, the "Is it?," "What kind is it?," and "Why is it?" by various kinds of argumentation, therefore [definition and argumentation suffice for replying to the four questions]; yet again, an instrument receives its name from its proximate end, but division is ordered directly to {definition},[9] method to order,[10] proposition to composition; therefore [there are no proximate ends from which these instruments can be denominated].

[4] Nor can you object: proposition is a third instrument distinct from argumentation and definition that by itself serves perfectly to direct the operations of the intellect. The reason is that the proposition, if considered as one of the premises, is ordered to the syllogism; if, on the other hand, it is considered as a principle that is self evident, then, since it is known by the natural light [of the intellect],[11] it does not make a

separate instrument; finally, if it is considered as one of the principles, but unknown, this would not make it a separate instrument either, since it would have to be proved through a demonstration, or through a syllogism, or through an induction. It is sufficient therefore that the proposition, as disposing the intellect for correct composition, be enumerated among the six instruments that serve mediately to direct the operations of the intellect.

[5] If we focus, on the other hand, on instruments that perfectly and immediately serve to direct the operations of the intellect, these are only two: definition and demonstration. For whatever is known perfectly is either the nature of a thing or some property of it; but a nature is known perfectly through definition, a property through demonstration; therefore [definition and demonstration alone suffice]. Nor can you say: induction and demonstration of the fact ought to make one instrument that is essentially distinct from these two. The reason is that knowledge had through induction and demonstration of the fact is *a posteriori* and imperfect; but here we are speaking of instruments that serve knowledge perfectly. Of these there are only definition and demonstration, for these enable us to know a thing *a priori* and through its intrinsic or extrinsic causes.[12]

[6] Note, fourth: although all instruments of logic have in common that they direct operations of the intellect,[13] that they are necessary for all the sciences, and that they assist one another, they are nonetheless different in value. For division is superior to method and proposition, because it serves the first operation of the intellect more than any other instrument; method is superior to proposition because it ranges through everything else; and discourse or syllogism is superior to proposition because it pertains to the third operation of the intellect, which takes precedence over the second operation. But since demonstration and definition serve to direct the operations of the intellect by themselves and immediately, and are so treated by the logician, they are the most important of the instruments. And the entire difficulty is this: which of these is superior?

[7] The first opinion is that of Simplicius and of Ammonius in the preface to the *Categories*, of Philoponus in the beginning of the *Prior Analytics* and on this first book of the *Posterior Analytics*, and of all the Latins, holding that demonstration is the most important of all logical instruments. They all offer the following arguments in support of their opinion:

[8] First: demonstration is the goal of all matters treated in logic; for the book of the *Categories* is ordered to the proposition; the proposition to the syllogism; and the syllogism to demonstration. But whatever is the goal in science is the most important [of its instruments]; therefore [demonstration is the most important instrument].

[9] Second: the superior instrument should serve the superior operation; but discourse is the best operation, and demonstration is its instrument; therefore [demonstration is the superior instrument]. Proof of the minor: because discourse is concerned with the superior object; also, because, although we might be similar to intelligences[14] in the first operation [of the intellect] with respect to the manner of functioning, we are more similar to intelligences with respect to the object, which is far superior, through the third operation [of the intellect], because man is constituted in his very being through discourse; but whatever is constitutive of a thing[15] is most important; [therefore discourse is the superior operation of the intellect].

[10] Third: the superior instrument is what makes us know a thing through a cause; but demonstration alone is of this kind, since Aristotle says that we think we know a thing scientifically when we believe that we have demonstrated something [of it through a cause]; therefore [demonstration is the superior instrument].

[11] Fourth: the superior instrument is what is concerned with [a] the superior and [b] the more extensive object[16]; but definition is concerned only with the quiddity[17] whereas demonstration is concerned with both the quiddities and the properties of things; therefore [demonstration is the superior instrument].

[12] Fifth: the whole is superior to its part and the container is superior to the thing contained, from Aristotle, the fourth book of *De caelo*, texts 25, 36, and 39; but definition is a part of, and is contained in, demonstration; therefore [demonstration is superior to definition]. Proof of the minor, from Aristotle in the first book of *De anima*, 11, and in the first book of the *Posterior Analytics*, text 24, saying that the "what it is"[18] is a principle in demonstration; but every principle is included in whatever makes it be a principle; therefore [demonstration is superior to definition].

[13] You say: this argument proves that definition is contained in demonstration as a middle term, not as an instrument of scientific knowing.[19] To the contrary: a definition is put in a demonstration so that, when the nature of the subject is known, we may gather one or other of its properties; therefore a definition is put in a demonstration as an

instrument of scientific knowing and not as a middle term. The antecedent is apparent from the following example: every rational animal is discursive; every man is a rational animal; therefore [every man is discursive].

[14] The second opinion is that of Averroes in the first and second books of the *Posterior Analytics* and in the *Epitome of Logic*, in the chapter on demonstration; of Balduinus, in his query on this matter; and of Mirandulanus and others, holding that definition is the best of instruments because definition is concerned with a better object than demonstration, because definition serves a more important operation, and because the most important demonstrations depend on definition for their certitude.

[15] The third opinion is that of many moderns[20] who, following a middle road, say: definition can be taken in two ways, either in itself and in its nature, or with respect to us[21] and as it is in us. Taken in the first way definition precedes demonstration in dignity, whereas taken in the second way it does not, because the third operation [of the intellect] is more perfect than the first considered as it is in us; also, because demonstrations bring certitude to definitions as they are in us, since the parts of a definition cannot be known except through demonstration.

[16] For the solution of this difficulty,[22] note first: definition can be taken in several ways: for the quiddity and nature of a thing; or for a topical instrument from which probable arguments can be drawn; or for the middle term of a demonstration; or, finally, for an instrument of scientific knowing. The problem concerns definition taken in the fourth way.

[17] Note, second: a science takes its quiddity from two things, namely, the importance of its object and its evidence; thus an instrument of a science should take its worth from these also. Proof: an instrument is specified by its object; therefore, the better the object the better the instrument. Second, just as the better the science the more certain and the more evident it will be, so the instrument that makes it be more certain and more evident will be the better instrument.

[18] Note, third: the first operation of the intellect is much superior to the third. First: it is concerned with a superior object, that is, with the quiddity and essence of a thing, whereas the second shows only that something exists in something else and the third knows the subject only as the cause of another; therefore [the first operation is superior to the other two]. Second: it is concerned with its object in a better way, that is,

without movement, and the better something is the closer will it approach the best; but the third operation involves movement[23]; therefore [the first operation is superior to the third]. Third: although definition depends on demonstration instrumentally, nonetheless definition is itself the goal of demonstration; but to be the goal of something is to precede it in dignity. Add to this: most powerful demonstrations depend on definition.

[19] Themistius adopted this opinion in the first book of *De anima*, chapter 24, and proved it with these arguments. First: by the first operation we either know something to be completely true or we do not know it at all; by the second and third operations we can know both truth and falsity. Second: to understand something is to embrace it and touch it in some way; but in the second and third operations we merely go around and circle the thing, whereas by the first we grasp it; therefore [the first operation is superior to the second and the third].

[20] Note, fourth: it is one thing to know a thing perfectly, another to know only its nature and some of its properties. For the first, definition along with demonstration is required; for the second, if the nature alone is known, definition suffices, if the properties, demonstration is enough.

[21] I say, first: definition in itself is far superior to demonstration. Proof, first: for definition makes substance known, demonstration only accident, from the second book of the *Posterior Analytics*, chapter 2; but it is better to know substance than accident; therefore [definition is superior to demonstration]. You say: both substance and accident are known through demonstration. To the contrary:

[22] Second: the premises of a demonstration cannot be known through the demonstration whose premises they are, but by some other instrument, and this is nothing but the definition, which generates knowledge of the quiddity that enters into the demonstration as a premise; therefore [the premises are known by definition and not directly by demonstration]. Confirmation: for each task there is only one natural and adequate instrument; but quiddity is known through definition; therefore it cannot be known through demonstration. Second: the instrument and the thing known [by it] are correlatives.[24] Third: otherwise one or the other would be superfluous. Fourth: quiddity is something simple,[25] and this can only be the object of the first operation of the intellect.

[23] Third: the end is superior to things ordered to it, from the fifth book of the *Metaphysics*, chapter 2; but definition is the end of demonstration; therefore [definition is superior to demonstration]. Proof of the minor, from Aristotle in the twelfth book of the *Metaphysics*, in the

beginning, where he says that accidents are ordered to substance as to an end. From this I argue as follows: accidents are ordered to substance as to an end; therefore the knowledge of those things that are attained through demonstration is also ordered to the knowledge of substance, and this is grasped through definition.

[24] Fourth: it is better to know the formal cause as such than as the cause of properties; but it is known as such through definition, as the cause of properties through demonstration; therefore [definition is preferable to demonstration]. Proof of the minor: definition produces knowledge of the quiddity precisely as quiddity, whereas demonstration manifests some property of a subject through a middle term. From this it follows that the argument sometimes made to the contrary is not valid, [namely,] that we assent to a demonstration once had, but not to a definition; therefore demonstration is preferable. First: we do not assent to a demonstration except in virtue of a definition. Second: demonstration is ordered to definition as to an end; and if, to the contrary, definition were ordered in some way to demonstration, this would be so as to communicate its own perfection to it.

[25] You object: the quiddity of a thing can be demonstrated through a final cause, as is obvious in this syllogism: Every living being procreated to attain beatitude[26] is a rational animal; but every man is such; therefore [every man is a rational animal]. I reply, first: this does not manifest a quiddity, because a quiddity is known through formal causes, whereas the foregoing demonstration argues from extrinsic causes.[27] Second: the quiddity in such a demonstration is shown materially, not formally; for, as Aristotle proves at length in the second book of the *Posterior Analytics*, quiddity precisely as quiddity cannot be demonstrated.

[26] I say, second: definition as it is in us is superior to demonstration. Proof of the conclusion: even if definition as it is in us were to depend on syllogism, nonetheless, since definition is the end of the syllogism and depends on it only instrumentally, [it is superior to demonstration or to syllogism]. Also, definition even as it is in us is concerned with a better object and in a better way than demonstration, that is, without {movement}.[28] Again, most powerful demonstrations as they are found in us depend on definition for their certitude. Yet again, nature always seeks what is best; but nature is more desirous of knowing the cause of a substance than of an accident; therefore [definition, which reveals the cause of a substance, is superior to demonstration, which reveals the cause of an accident].

[27] You object: we {barely}[29] are able to know any quiddity; therefore definition as it is in us will not be a better instrument than demonstration. Reply: the same would have to be said of demonstration, for this supposes not only definition but other conditions as well. Second: here teachings on definition and demonstration are intertwined, and thus the argument is not to the point.

[28] You object, second: definition and the first operation of the intellect as it is in us do not seem to offer more certitude than do the third operation and demonstration; therefore definition as it is in us will not be superior to demonstration. I reply, first: definition and the first operation have their own certitude independently of that of demonstration; for, if this is known, namely, that knowledge of the genus and differentia of a thing being posited the knowledge of the quiddity and nature of that thing is also immediately posited, it is likewise the case that whoever first recognizes that this is the genus and this the differentia of the thing knows the quiddity of that thing independently of demonstration. I reply, second: even if definition as it is in us depends on demonstration instrumentally – just as intellectual knowledge depends on sense knowledge and demonstration of the reasoned fact depends on demonstration of the fact instrumentally – no one would venture to hold that demonstration of the reasoned fact or intellectual knowledge is not superior to demonstration of the fact or sense knowledge; therefore neither should one hold that definition is less perfect than demonstration.

[29] From these considerations I infer, first: definition and demonstration are related analogically as instruments of scientific knowing, in such a way, however, that definition is the primary analogate; they are analogous by an analogy of proportion and also of attribution[30]; of proportion, because just as definition is related to knowledge of quiddity so demonstration is related to knowledge of properties; of attribution, because, since definition is the end of demonstration, and demonstration, especially most powerful demonstration, depends on definition for its certitude, [definition is the primary analogate to which demonstration is attributed as an instrument of scientific knowing].

[30] I gather, second: definition is the end of demonstration; because everything imperfect is reducible to the perfect; but demonstration is imperfect when compared with definition; therefore [definition is more perfect than demonstration and its end]. Also, just as accidents are for substance, from Aristotle as above [23], so also is the knowledge of accidents for the knowledge of substance itself.

[31] The reply to the first objection [8] is obvious from the foregoing.

[32] To the second [9], I reply: the first operation is superior to the third, as we have shown in the last instance [30]; for the other difficulties the reply is obvious from the foregoing. I reply, [second]: man differs from other beings through the first operation in a way superior to that through the third, since the first operation is superior.

[33] To the third[31] [11], I reply, concerning the first part [11b]: it is more obvious that demonstration is an instrument for knowing accidents. Second: even accidents can be known through definition, since they can be defined in their own way, indeed in a better way, for it is more important to know what they are than why they exist in substances. I reply, second, concerning the second part [11a]: definition is concerned with the superior object, as we have shown above [26].

[34] To the fourth[32] objection [10], I reply: universally speaking the major premise is false, because first principles make us know, and not through a cause as does demonstration; moreover first principles are superior because they are the efficient cause of demonstrative science; therefore [definition is superior to demonstration]. I reply, second, by denying the minor premise, because definition makes us know through intrinsic and formal causes, and these are the most perfect of all.

[35] To the fifth[33] [12], I reply: definition as it is an instrument of scientific knowing is not a part of demonstration and is not contained within it.

[36] To the sixth,[34] I reply: definition as it occurs in demonstration is ordered to the conclusion as an end and terminus, just as are the premises; indeed, because in a demonstration definition is ordered to the conclusion only as it is the cause of proper attributes, it is not a part of demonstration as an instrument of scientific knowing. I reply, second: the quiddity and nature of a thing is indeed known in a demonstration, but through the definition itself as it is an instrument of scientific knowing.

[37] You say[35]: the entire certitude of definition, to say the least, depends on demonstration, for otherwise there would be no need for the type of demonstration wherein the parts of a definition are discovered. This is obvious from the example of other instruments, which are no longer necessary after the task to which they are essentially ordained is completed; but demonstration is always necessary for definition so that the definition will be grasped with certitude; therefore [definition is always dependent on demonstration]. I reply[36]: there is a difference between instruments of scientific knowing and other instruments, because

we cannot know scientifically unless we know that we know with certitude; from this it follows that instruments of scientific knowing are always necessary, at least virtually, for perfect knowing. This is apparent from the example of demonstration: for one cannot know something through demonstration unless one has a knowledge of demonstration in general. The same does not apply to instruments serving the other arts.

[D2] *Second Disputation: On the Properties[1] of Demonstration*

[D2.1] *First Question: Is demonstration composed of true premises[2]?*

[1] In this question we will solve three difficulties.[3] The first is: when Aristotle says that the principles of demonstration must be true, of what truth is he speaking, complex or non-complex?

[2] Note that truth is twofold[4]: one is real[5] [i.e., ontological truth], which is nothing more than the conformity of a thing with its first real principles; the other is of reason[6] [i.e., epistemological truth], which is found only in a knowing power. The latter is also twofold: one is simple,[7] of the kind found in the first operation[8] of the intellect and in sense knowledge, of which Aristotle is mindful in the second book of *De anima*, 66; and this is nothing more than the conformity of what is apprehended with the thing that is outside the knower. The other [is] complex,[7] and this is found in both the second and the third operation[8] of the intellect, and it is nothing more than the conformity of the proposition made by the intellect with the unity that is outside the knower. On this account the following proposition is said to be true: "Man is a rational, risible animal," because, on the part of the object outside, man is truly conjoined with animality, [rationality, and risibility].

[3] This understood, I say: when Aristotle says that principles must be true he is speaking of complex truth, because the principles of demonstration can have only complex truth, and because otherwise Aristotle would have proved nothing. For, he says, demonstration must be made from true principles because what is known is true; but a conclusion is true with complex truth; therefore [the principles on which it is based must be true with complex truth also].

[4] First objection: truth can follow from false premises; therefore demonstration does not necessarily contain true premises. I reply, first, with Averroes: truth follows from false premises accidentally and not essentially[9]; but demonstration must be made from principles that are

related essentially[10]; therefore [the principles of demonstration must be true]. I reply, second, with St. Thomas: truth can indeed be inferred from false premises, but it cannot be known scientifically[11] from them, because over and above the inference, for something to be known scientifically there must also be proof,[12] and proof cannot exist unless there are true premises.

[5] Second objection: demonstration to the impossible is true demonstration and yet it is made from false premises; therefore [the premises of a demonstration need not be true]. I reply: here we are discussing ostensive demonstration[13]; therefore [this peculiar feature of demonstration to the impossible does not apply here, since the premises of ostensive demonstation must be true]. Second: just as demonstration to the impossible is demonstration in a qualified sense,[14] so it is also based on principles in a qualified sense, that is, principles directed to the person[15] [and not to the issue].

[6] The second difficulty: does demonstration require both complex and non-complex[16] truth?

[7] I say, first: on the part of its principles demonstration requires only complex truth, because this is all it is capable of. I say, second: on the part of its object it also requires non-complex truth. The proof: from Aristotle, who clearly indicates this when he teaches that demonstration must be based on causes and that science is knowledge of the cause making the thing be what it is. Second: the intellect, like each of the other powers, is perfected in relation to its proper object; but the proper object of the intellect is real being,[17] which is true by non-complex truth; therefore demonstration as an instrument perfecting the intellect must be concerned [with an object that is true by non-complex truth].

[8] From this it follows, first: concerning the vacuum, the infinite, and similar things one can form propositions that are true by complex truth, but there cannot be science of them, because they do not exist. It follows, second: there cannot be a science of rational being,[18] because a science as a perfection of the intellect must be concerned with real being. It follows, third: falsehood cannot be known scientifically, for otherwise demonstration would not be based on true premises.

[9] The third and last difficulty: what does Aristotle's proposition, "What is not, is not known," mean? Cajetan notes correctly that this proposition can have two senses. One is: what is not, i.e., what does not exist, is not known. This sense is false, because there can be science even of non-existent things, as is apparent with a rose [in winter][19]; and hence

this sense was not intended by Aristotle. The other is: what is not, is not known, i.e., what is not in a thing cannot be known [to be in the thing].[20] And this sense is true; indeed it is true because otherwise falsehood could be inferred from true premises, which is absurd. It is the sense intended by Aristotle, who states that demonstration must be established from true premises because the conclusion that is known is true.

[D2.2] *Second Question: Must demonstration be made from premises that are first and prior[1]?*

[1] These two properties, though formally different from each other,[2] are so interrelated that whatever are causes [that is, whatever are first] are also prior, and therefore they are treated together. The reason for raising the question: [a] causes are more known only with respect to nature, but demonstration must also be based on things more known with respect to us[3]; therefore [the question arises: in what sense can the premises of a demonstration be said to be first and prior?]. Also, [b] one can have a most perfect science[4] of God and yet there are no causes in him[5]; therefore [it would seem that demonstrations need not be made from premises that are first and prior the way causes are].

[2] Note, first: causes in being,[6] as Cajetan correctly teaches, are of two kinds: some are true and proper causes, which produce an effect that is distinct from themselves at least formally; such a cause is rational animal with respect to its attributes and properties. Others are imperfect[7] and do not produce an effect and are not formally distinct from what they achieve. They are said to be causes because, if a thing could have a cause for the attributes that are predicated of it, these would be those causes. I explain: since God is most simple act, there can be nothing in him that has the formality of cause; but if there could be something in him having the formality of cause, immutability would be the cause of eternity, immateriality would be the cause of impassibility, and justice would be the cause of retribution. Nor can you say: this distinction was unknown to Aristotle and is not in accord with his teaching. For in the second book of *De caelo*, text 13, Aristotle teaches that the motion of the heavens began in the east; then he objects against himself that the motion of the heavens had no beginning; therefore [it could not have begun in the east]. He replies: the motion of the heavens began in the east, not because it truly did begin there, but because, if it had a beginning, it would have been there. And this is nothing more than

saying that the eastern region is the virtual beginning of the motion of the heavens.

[3] Note, second: this question depends on two others. The first: can the demonstration of which we speak be made through all [four] causes? The second: for such a demonstration is it necessary to have causes that are true and proper in the order of being, or do causes that are merely virtual suffice? Cajetan holds the second whereas others do not.

[4] I say, first: demonstration of the reasoned fact can be made from all species of cause,[8] but preferably it should be made from the formal cause. Proof of the first part, from Aristotle, who offers extensive proof of this in the second book of the *Posterior Analytics*, and from experience: for sometimes we demonstrate a conclusion from a final, sometimes from an efficient, sometimes from a formal, and sometimes from a material cause. Proof of the second part: scientific knowing is the effect of demonstration, and it consists in knowing a thing through the cause that makes it be what it is; therefore a demonstration will be the more perfect the more it proceeds from formal causes, which are more intrinsic to the thing.[9]

[5] I say, second: there can be demonstration of the reasoned fact only through causes that are true and proper in the order of being. For to know scientifically is to know a thing through a cause that makes it be what it is; but only causes that are true and proper in the order of being make a thing be what it is; therefore only through them can demonstration of the reasoned fact be made. Second: demonstration of the reasoned fact produces scientific knowing of the thing in an unqualified way; but demonstration that proceeds from true causes effects this, whereas the kind that proceeds from virtual causes, being made from a supposition,[10] is unable to produce such knowledge in an unqualified way.

[6] To the first objection [1a] I reply: something can be more known, in the matter that concerns us, in two ways: either with respect to nature and with respect to us, and in this way the first principles of mathematics[11] are more known, and these offer no difficulty; or with respect to nature alone, and then, since nothing can be proved except through something that is more known, a cause must be manifest before it can be taken as a premise in a demonstration. And from this the reply to the objection is obvious. To the second objection [1b] I reply: since God does not have a genus or a difference, and since nothing can be designated in him that can have the formality of being causative of him, it follows that there can be neither demonstration nor definition [of him], nor, as a consequence,

science. But there still can be a certain higher kind of knowledge of him.[12]

[7] I say, third: demonstration making use of causes that are virtual in the order of being is imperfect and verisimilar. Proof of the conclusion: first, since things are related to being as they are related to knowing, a thing with causes that are imperfect may be known through them. Nor can you say that these virtual causes {are}[13] merely principles of knowing, and therefore any demonstration that proceeds *a posteriori* can be said to be imperfect and verisimilar. Virtual causes are principles not merely in the order of knowing but also in the order of being[14]; still they are not proper and perfect, but imperfect and verisimilar.

[8] You ask, first: did Aristotle include demonstration based on virtual causes in his definition of demonstration? I reply: properly speaking, no; proportionally, yes.

[9] You ask, second: why must demonstration be made from causes that are true and proper in the order of being? I reply: the middle term must supply the reason for connecting the major term with the minor term and proving the major of the minor; but this cannot be done unless the demonstration is based on true causes; therefore [demonstration must be based on causes that are true and proper in the order of being]. Second: as a thing is related to being [so it is related to knowing], and since it is dependent on its causes for its being it must also be known through them.

[10] You say: demonstration can be made from a definition, and yet a definition is not a cause; therefore [demonstration need not be made from causes]. I reply, first: a definition implicitly involves formal causes.[15] I reply, second: a definition, as many hold, is the equivalent of either an efficient or a formal cause.

[11] You ask, third: must a demonstration always be made from proximate causes? The reason for raising the question is that the following demonstration is universally regarded as optimal: "Every rational animal is risible; every man is a rational animal; therefore every man is risible"; nonetheless it is not based on causes that are immediate because the proximate cause of risibility is the ability to wonder, not the ability to reason. The reply: demonstration of the reasoned fact must always be made from causes that are proximate, either actually or virtually.[16] To the objection I reply: "rational animal" can be taken in two ways in the foregoing demonstration: either as it implies only the indicated grades of definition,[17] and in this way it cannot be used in a demonstration; or, in a second way, as it contains both "discursive" and "wondering" virtually,

and then it can be used in a demonstration and it is a virtual cause proximate to the property being demonstrated.

[D2.3] *Third Question: What does Aristotle mean by immediate propositions*[1] *when he teaches that demonstration must be made from them?*

[1] To understand this question, note, first: from Averroes in many places and from *De primitate praedicatorum*, "first" can be taken in two ways: either for first to a subject or for first to a cause. First to a subject is said of a proposition in which the predicate is predicated of the subject immediately; first to a cause is said of a proposition in which the predicate is said of the subject with no other cause coming in between. Thus the proposition "Man is risible" is first to a subject because risibility is predicated of man as of a subject in such a way that no other subject comes in between. But the proposition "Man is a rational animal" is first to a cause, because no cause comes between rational animal and man through which rational animal can be proved of man. From this it follows that what is immediate to a subject {need}[2] not be first to a cause, but what is first to a cause is always first to a subject.

[2] Note, second: first to a cause can be understood in two ways: either for a proposition that is first and immediate with respect to a conclusion, or for a proposition that has no cause above it through which it can be proved *a priori*.

[3] I say, first: by an immediate proposition Aristotle does not mean a proposition that is first to a subject or first with respect to a conclusion. Proof: because propositions of these types[3] can often be demonstrated *a priori* through another proposition.

[4] I say, second: by immediate propositions Aristotle means propositions with no others over them through which they can be demonstrated *a priori* and in the same genus of cause, or, those that have no cause between their subject and predicate through which [they can be proved *a priori*]. Explanation and proof of the conclusion: I say "proved *a priori*," because, although immediate propositions have no proposition through which they can be proved in the strict sense, nothing prevents their having axioms over them through which they can be proved and manifested in an improper sense. This is clear from the following example: for the proposition "Man is a rational animal" has no proposition over it through which it can be proved in a strict sense, but it

does have axioms through which it can be demonstrated in an improper sense,[4] such as "Contradictory propositions cannot both be true at the same time." I say second, "*a priori*," because nothing prevents an immediate proposition from having another proposition over it, provided that the proof is not *a priori*; for *a posteriori* and with respect to our knowledge there can be a proposition through which an immediate proposition is proved; thus, for example, "Man is a rational animal" can be proved *a posteriori* from certain of his properties.[5] I say, third, "in the same genus of cause," because nothing prevents an immediate proposition from being proved in a different genus, as is apparent in the example, "Every living being procreated for beatitude[6] is a rational animal; but every man is an animal procreated for beatitude; therefore [every man is a rational animal]." In this example his being a rational animal is indeed demonstrated of man *a priori*, but through the final cause; therefore [nothing prevents an immediate proposition from being demonstrated *a priori* in a different genus of cause].

[5] You ask: is a proposition in which a primary property is predicated of its adequate subject immediate? For Cajetan affirms this, because no cause can be found through which it may be proved *a priori*. This is obvious, for, if there were any, it would be the definition; but a definition is not distinct from the thing defined; and, if it were distinct in any way, it would not serve as a cause through which such a proposition would be manifested *a priori*; therefore [no cause can be found through which it may be proved *a priori*]. I reply: a proposition in which a property is predicated [of its adequate subject] is not immediate, because it does have a proposition over it through which it can be demonstrated, namely, the definition. My reply to Cajetan's argument: the thing defined can be taken in two ways: either as the thing defined, and in this way it is the same as the definition "from the nature of the thing"[7]; or second, as it includes existence and subsistence, which are modes from which the definition abstracts, and on this account differs from the definition formally and the definition can be the cause that explains why the property is present in it. You say, for Cajetan: whenever anything is demonstrated of the thing defined through the definition this begs the question,[8] which he seems to teach along with Thomas, Albert, and Aristotle. My reply: whenever anything is proved of the thing defined through a definition that is equally unknown, this begs the question, and that is what the aforementioned authors meant; but when the definition is clearer [than the thing defined], the question is not begged and something can be proved of the thing defined.

[6] You ask, second: what are these immediate propositions? My reply: they are, first, those in which a definition or a part of a definition is predicated of the thing defined; second, those in which a primary property is predicated of the definition of the subject; third, those in which attributes are predicated of God; fourth, those in which one category [of being] is denied of another[9]; fifth, those in which one differentia is denied of another.[10]

[7] Note here that Aristotle divides immediate propositions otherwise,[11] i.e., into axioms and positions. Axioms are propositions that must be known by anyone who would learn a science. Positions are propositions that need not be foreknown by one learning a science. These are twofold: some are suppositions,[12] in which one thing is affirmed or denied of another; others are {terms} or definitions,[13] in which nothing is affirmed or denied of another.

[8] You ask, third: must every immediate proposition be self evident[14]? My reply: propositions are known in two ways, either with respect to nature or with respect to us. Those known with respect to us are those that, from the viewpoint of our knowledge, have no proposition over them through which they can be proved; those known with respect to nature are those that, from the nature of the thing,[15] have no proposition over them through which they can be proved.

[9] I say, first: these matters understood, every immediate proposition is known with respect to nature, for otherwise it would not be immediate, since it would have another proposition over it through which it could be proved.

[10] I say, second: not every immediate proposition is known with respect to us; for only axioms are of this kind, since they, being most known, have no proposition over them that is more known and through which [they can be proved].

[D2.4] *Fourth Question: Must every demonstration be made from immediate premises,[1] and how?*

[1] It seems not: [a] for otherwise for each thesis there would be a single demonstration that is most powerful,[2] namely, that in which a proper attribute[3] is proved of the subject through its definition; furthermore, [b] because it seems no less characteristic[4] of a demonstration to be made from causes than to be made from immediate premises, and a single cause alone suffices for a perfect demonstration, therefore a single immediate premise should suffice[5] for it also.

[2] I say, first: every demonstration must be made in some way[6] from immediate premises. Proof: from the authority of Aristotle in this second chapter [of the *Posterior Analytics*], and from reason: for to know scientifically is nothing other than to assent to a conclusion with certitude and evidence; but one cannot assent to a conclusion with certitude and evidence if the demonstration is not made from immediate premises; therefore every demonstration must be made from immediate premises. You say: a subalternated science has perfect demonstrations, and yet it supposes its immediate principles as proved in its {subalternating} science[7]; therefore [a perfect demonstration need not be based on immediate principles]. I reply: a subalternated science is imperfect and does not have perfect demonstrations, since it presupposes its first principles as proved in the higher science; thus it generates a science from a supposition and in a qualified way.[8] From this the solution is obvious.

[3] I say, second: a most powerful demonstration must be made from indemonstrables,[9] and this is proved both from the authority of Aristotle and from reason. For scientific knowing is the more perfect the more certain it is; but scientific knowing that is had through a demonstration from indemonstrables is most certain; therefore a demonstration based on indemonstrables, generating as it does most perfect science, will be most perfect.[10]

[4] I say, third: a demonstration based on virtually indemonstrable premises[11] is true demonstration, though less perfect than the foregoing. Proof of the first part: from Aristotle, in the first book of the *Topics*, chapter 1, saying that demonstration is made either from true and first premises, or at least from premises whose knowledge originates from true and first premises – which is to say that demonstration is based on premises that are immediate either actually or virtually.[12] Also, from reason: a demonstration based on causes produces true scientific knowing and is true demonstration; but a demonstration based on premises that are virtually immediate is of this kind; therefore [a demonstration consisting of premises that are virtually indemonstrable is true demonstration]. Proof of the second part of the conclusion: a demonstration made from actually indemonstrable premises is independent, whereas one made from premises that are virtually immediate is dependent on another; but it is more perfect for a demonstration to be independent than for it to depend on another; therefore [a demonstration made from premises that are virtually immediate is less perfect than one made from those that are actually immediate].

[5] You ask: does Aristotle include these demonstrations in his definition? I reply affirmatively, with St. Thomas and Philoponus and against Averroes. The reason: because demonstration in general, taken analogically,[13] is said of demonstration made from premises that are indemonstrable either actually or virtually; but Aristotle usually treats analogates together; therefore [he would include these types of demonstration under his general definition]. That the major premise is true is obvious, because a demonstration made from premises that are virtually indemonstrable produces scientific knowing only when complemented by a demonstration made from premises that are actually immediate.

[6] To the first objection [1a]: I reply by denying the inference, because if an addition that serves as a connector[14] is made next to the middle term of a demonstration, or indeed next to the subject, there can result a large number of most powerful demonstrations. This is apparent if one were to say: "Every discursive animal is capable of wonder; every man [is a discursive animal]; therefore [every man is capable of wonder]"; or, "Every being capable of wonder is risible[15]; but every man as a rational animal is capable of wonder; therefore [every man is risible]." But note here that by a most powerful demonstration we mean one in which a property is demonstrated of a subject through the {subject's} definition.[16]

[7] To the second objection [1b]: I reply by denying the truth of the argument: because, when a thing has only one cause requisite for it to be known perfectly through a demonstration, knowledge of that single cause suffices; yet, since a demonstration depends on both premises, each of these must be immediate for the demonstration to be perfect.

[D2.5] *Fifth Question*[1]: *Do all immediate self-evident principles enter into each demonstration?*

[1] It is certain that principles of this type do not enter actually into each demonstration, and all are agreed on this. The problem is whether they enter virtually, because of Aristotle's teaching in the first text that first principles must be foreknown to be true, and Philoponus's taking the term for such principles to include even axioms.[2]

[2] The first opinion is that of Giles, Apollinaris, and Paul of Venice, who affirm this, first, [a] because Aristotle teaches in the first text of this book that first principles must be foreknown to be true[3] and also because they enter the demonstration virtually[4]; therefore [this is how he is to be

understood]. Second, [b] because of Aristotle's teaching in text 25[5] of the same book that science is concerned with genus, properties, and axioms; therefore [axioms enter into demonstrations]. Third, [c] because all syllogisms depend on these two principles: "Whatever can be affirmed of a totality can be affirmed of the members included under it" and "Whatever can be denied of a totality can be denied of the members included under it"; therefore [these principles enter into every syllogism].

[3] The second opinion, which is almost universally common, denies this. The basis: because only principles on which the conclusion depends intrinsically enter into its demonstration actually or habitually; but a conclusion depends intrinsically only on principles that are proper and intrinsic; therefore it does not depend on those that are self evident, since these are extrinsic.

[4] Note, first: the term axiom can be taken in two ways, either properly and strictly, or improperly and broadly. Taken in the first way it includes only propositions that are most known and common to all or some of the sciences, of which kind is "Every whole is greater than its part."[6] In the second way it includes not only such propositions but also all principles that are proper, immediate, etc.,[7] and it is in this way that it was taken by Aristotle in the same book, texts 20 and 25.[8]

[5] Note, second: the term axiom taken in this second way can be further understood in two ways: either as it is common to all the sciences[9]; or only to some, as it is restricted to some determinate subject matter,[10] as is apparent from the example of "Contradictories cannot both be true at the same time." This principle can be considered either in general, as applicable in all the sciences, or as it already has been actually applied to a particular science, e.g., mathematics.[11]

[6] These points understood, I say, first: axioms that are immediately self evident can enter into an imperfect demonstration[12] actually, as they can also enter into an improper one; so Philoponus and Themistius. Proof, from Aristotle, texts 24 and 26[13] of the first book of the *Posterior Analytics*, who expressly teaches this, particularly of demonstration that reduces to the impossible. Also, from reason: a principle that is immediate and self evident can be put in place of a premise, as is done in mathematical demonstrations[14]; but a principle enters into a demonstration actually when the demonstration is actually made from it; therefore [an axiom can enter into a demonstration actually]. Nor can you say: a demonstration must be made from causes that are true and immediate; but the aforementioned axioms are not of this type; therefore

[they cannot enter a demonstration actually]. For this proves only that axioms cannot enter into a perfect and proper demonstration, not into one that is improper and imperfect; and, indeed, when such axioms are restricted to a particular subject matter (and only in this way do they actually enter a demonstration) they have a great similarity to true causes, since they are prior, more known, simpler, and causes in relation to us.

[7] I say, second: such axioms[15] do not enter any demonstration[16] either actually or virtually. Proof of the conclusion: only principles on which the conclusion depends intrinsically enter into the demonstration actually or virtually; but a conclusion cannot depend intrinsically on such axioms, since they are extrinsic principles; therefore [they do not enter actually or virtually into a demonstration]. From which it follows: the afore-mentioned axioms do not enter even virtually into mathematical demonstrations if the demonstrations are considered in themselves, although they do enter into them if such demonstrations are considered in relation to our knowledge,[17] because the truth of such axioms is more known to us than are the proper principles in a mathematical demon-stration.

[8] To the first objection [2a], I reply: Aristotle is speaking there of the foreknowledges required for any teaching or discipline, not merely those for demonstrative science. I reply, second: sometimes first principles must be foreknown to be true, namely, when such principles enter actually into {an imperfect} demonstration.[18]

[9] To the second [2b], I reply: Aristotle there took the term axiom in the second way, that is, most broadly as it includes all principles, proper and intrinsic as well as improper and extrinsic.

[10] To the third [2c], I reply by denying that every syllogism depends on those two principles: for they depend intrinsically only on their own proper principles. And if one says that they do depend on them, this is in the sense that one uses them against a person who denies the proper principles of a particular science[19] by arguing from these two principles as from something more known.

[11] Here one may inquire[20] whether or not a conclusion must be resolved to first principles[21] if one is to have perfect scientific knowledge of it. This supposed, scientific knowledge of a particular matter can be had in two ways: either in an unqualified way and absolutely, or in a qualified way and in a determinate genus.[22]

[12] I say, first[23]: for anyone to have perfect knowledge of any subject matter one must make resolution to all principles and causes, including

the first and most universal; or, better, one must know all principles and all causes, including the first. This is against Giles, who argues that resolution must be made to something more known; but the first cause is not of this type, and so it follows that a subject matter cannot be properly resolved to the first cause. Proof of the conclusion: just as things are related to being so they are related to knowing; but a conclusion depends for its being on a subject and a predicate, and these depend for their being both on their principles and on God; therefore [perfect knowledge of a subject matter requires resolution to all principles and causes, including the first].

[13] I say, second[24]: for a subject matter to be known perfectly in its genus it suffices to know its proper causes. Proof of the conclusion[25]: a thing depends on determinate causes for its determinate being, and if these are known perfectly, then the thing itself is known in its genus. But with respect to us,[26] for us to know perfectly the conclusion must be resolved to first principles that are self evident, since with respect to us such principles are also most known; and against those who deny them the refutation must be based on those principles.

[D2.6] *Sixth Question[1]: Is demonstration made from premises that are more known,[2] and is knowledge of the premises[3] better and more perfect than[4] that of the conclusion?*

[1] Concerning the first query, note, first: "more known" has two senses, one with respect to nature, the other with respect to us.[5] More known with respect to nature is whatever is intelligible by a priority of nature, and this is said of things destined to have their existence earlier. More known with respect to us is whatever comes to be known first by us: an example would be the singular thing, because we perceive it by the senses and what is sensible is most known to us; again, the more known with respect to us is the more universal, because it is more connected with things we perceive by the senses by reason of accidents that are most known, such as quality, quantity, figure, and so on, which we also perceive by the senses, and from these we generally come to the knowledge of what is less universal. I said, "for the most part," for sometimes what is more universal is not more known to us, as can be seen of many things in the category of relation.

[2] But note here: when I say that what is more universal is more known to us, I take this to apply to confused knowledge, because if we are

speaking of distinct knowledge[6] what is more universal will be less known, because the more universal cannot be known distinctly unless the less universal is foreknown. You object: Aristotle teaches at the beginning of the *Physics*, text 4, that universals are more known to us than are singulars. By way of answer note that "singular" can be taken in two ways: either properly for a material individual; or improperly, for something that is less universal in relation to the more universal. Note, second: universals are of two kinds, either in the order of causality, as are the first mover and protomatter; or in the order of predication.[7]

[3] These distinctions understood,[8] I reply, first: singulars are more known to us than universals in the order of causality, and this Aristotle teaches in the first book of the *Posterior Analytics*, text 5.[9] I reply, second: less universals are less known to us than more universals [in the order of predication],[10] and this is the way Aristotle understood "less universals" in the first book of the *Physics*.[11]

[4] Note, second: following Averroes, most powerful demonstration must be made from premises that are more known with respect to us {and}[12] with respect to nature; following Philoponus, Themistius, and St. Thomas, on the other hand, they can be made from premises that are more known only with respect to nature. I think that the latter position is more probable, because most powerful demonstration must be made from proper and true causes, and these by their very essence are more known only with respect to nature. You object: before we can infer a conclusion from premises we must assent to them; therefore the premises must be more known to us; therefore [the premises must be more known both with respect to nature and with respect to us]. I reply: if the premises are considered in themselves, most powerful demonstration must be made from premises that are more known with respect to nature; if they are considered as taken by us to prove something else, they {indeed}[13] must be more known [to us][14]; but this does not count against our position.

[5] You say: can there be propositions that are known both with respect to nature and with respect to us? It would seem not, because any one thing precisely as one cannot be known both to us and to nature. I reply: there can be. This is apparent in mathematical demonstrations, where causes are more known than effects both to us and to nature, although demonstrations of this kind are not most powerful.[15] Also, what is more universal is more known to us than what is less universal, and since what is more universal is prior and simpler it is more known also with respect

to nature; therefore [what is more universal is more known both with respect to us and with respect to nature].

[6] Concerning the second query,[16] it seems that knowledge of the premises is not [better and more perfect than that of the conclusion]. First, [a] from experience: we assent no less to the proposition "Man is risible" than we do to "Man is a rational animal," and yet we prove man to be risible through his being a rational animal; therefore [knowledge of the premise is not better and more perfect than that of the conclusion]. Second: [b] if knowledge of the conclusion were less perfect, it would follow that a conclusion that depends on many principles would be very imperfect, and this goes contrary to the nature of science. Third: [c] we assent to knowledge of the conclusion, not to that of the principles; therefore [knowledge of the premises is not better than that of the conclusion]. Confirmation: [d] knowledge that is had through a cause is superior to that which is not; but knowledge of the conclusion is had through a cause, namely, through principles, whereas knowledge of principles is not; therefore [knowledge of principles is not superior to that of the conclusion]. Second confirmation: [e] knowledge that is more universal and more confused is more imperfect, and knowledge of principles is of this kind whereas that of the conclusion is not; therefore [knowledge of principles is not more perfect than that of the conclusion]. Fourth: [f] the end is superior to things ordered to the end; but knowledge of the conclusion is ordered to knowledge of principles; therefore [knowledge of the conclusion is superior to that of principles]. Fifth: [g] a natural cause that is not impeded[17] produces an effect equal to it in perfection; but knowledge of principles is the cause of natural knowledge of the conclusion and it is not impeded; therefore [knowledge of principles is equal in perfection to that of the conclusion]. Sixth: [h] if the knowledge of principles is greater than that of the conclusion, for example, if it were [measured] at three and knowledge of the conclusion at two, it would follow that, if each [degree of] knowledge[18] were gradually decreased, there would finally remain some knowledge of principles and none of the conclusion; but this is absurd, since a principle is only a principle with respect to the thing of which it is a principle.

[7] Note, first: the knowledge of premises is here being compared to knowledge of the conclusion insofar as the latter depends on the former. But from another viewpoint, for example, from sense knowledge or from another principle, the conclusion can be more known than the premises, as, for example, when known through sense knowledge or through faith, etc.

[8] Note, second: one knowledge is superior[19] to another, first, on the part of the object: for this reason Aristotle says rightly that it is much better to have slight knowledge of things divine with probability than it is to have scientific knowledge of things here below. Second: from the viewpoint of independence, for the more independent a science is, the better it is, as is apparent in subalternating sciences. Third: from the viewpoint of certitude, which consists in the firm adherence of the intellect to its object. [Fourth][20]: from the viewpoint of evidence, which is nothing more than a certain clarity in apprehending an object that makes the thing apprehended be more clearly perceived by us. Such evidence is twofold, either intuitive or discursive. Intuitive evidence is had through the knowledge of terms alone, and the evidence of first principles is of this kind; discursive evidence is had through causes, and such is the evidence of demonstrative science. Note here: although evidence is always accompanied by certitude – for we cannot assent to something with evidence if we do not firmly adhere to it – there can be certitude without evidence, as is apparent in subalternated sciences and more clearly in our faith.[21] Fifth: from the manner of knowing, for a science that is had through a more perfect cause is superior to one that is had through a less perfect cause.

[9] I say, first: knowledge of first principles[22] is more evident than knowledge of the conclusion. Proof of the conclusion: because knowledge of such principles is simpler and prior and functions as the cause of the conclusion deduced from them. Again, knowledge of first principles, being had through knowledge of terms alone, is so evident that it can easily be grasped by those who are inexperienced, whereas knowledge of a conclusion deduced from first principles is not so evident, for the intellect grasps things by themselves with greater evidence than it does things that depend on others, and a conclusion is of the latter type since it depends on first principles. The reason for this is that the intellect is more distracted the more it considers, and the less it considers the more clearly it understands. Thus knowledge of first principles [is called][23] wisdom by Aristotle, as opposed to that of the conclusion, which he calls science.

[10] I say, second: knowledge of immediate principles[24] is superior to that of the conclusion. Proof, from Aristotle, text 5 of this book, saying "that on account which something is, is even more so itself,"[25] and because one knowing principles would be better prepared than one knowing a conclusion; also because knowledge of immediate principles is

more independent and better. It is more independent, because knowledge of first principles does not depend on knowledge of a conclusion in the way knowledge of a conclusion depends on that of principles; better, because it is concerned with a superior object, namely, the cause of a property, and this is superior to the property itself, which is known only through a conclusion.

[11] To the first objection [6a], I reply: we give more assent to knowledge of the premises than we do to that of the conclusion, for the reasons given; however, sometimes because of the slight regress[26] we are not able to recognize this, as in the example given.

[12] To the second [6b], I reply: such knowledge is indeed very imperfect, but still it falls within the limits of science.

[13] To the third [6c], I reply: we give assent to knowledge of the conclusion, but we do so on account of our knowledge of first principles; therefore [knowledge of principles is superior to that of the conclusion]. To the first confirmation [6d], I reply by denying what has been assumed. To the second [6e], the answer is obvious from the foregoing: for, if we are speaking of knowledge of first principles, that is, of most universal principles, we may grant that from the viewpoint of the object such knowledge is more imperfect, since such principles are concerned with matters that are extrinsic and most common; but if we are speaking of knowledge of immediate principles, this is better, for the reasons given.

[14] To the fourth [6f], I reply: knowledge of the conclusion is superior to knowledge of the principles[27] in a qualified way, in the sense that knowledge of the premises is ordered in some way to knowledge of the conclusion, though it can also be denied that knowledge of the premises is ordered to knowledge of the conclusion essentially and primarily, since it is so ordered only secondarily. Absolutely speaking, however, knowledge of the premises is superior to knowledge of the conclusion, because the former is the efficient cause[28] of the latter.

[15] To the fifth [6g], I reply: the major premise[29] is true of a cause that is univocal[30] and equally perfect. Whence it follows, since knowledge of the premises is different in kind from knowledge of the conclusion, being more perfect and better, [that the conclusion deduced from that premise is not valid]. You say: at least knowledge of mediate principles, since this is similar in kind to that of the conclusion, will be no more perfect than knowledge of the conclusion. I deny the inference, because we cannot assent to a conclusion through mediate principles unless we do so in virtue of those that are immediate.

[16] To the sixth [6h], I reply: to the decrease of one degree in knowledge of the conclusion let there correspond a decrease of one and a half in knowledge of the premises, just as to the increase of one degree in knowledge of the conclusion there was needed one and a half degree in knowledge of the premises; in this way everything will correspond. You object: one makes progress from the more imperfect to the more perfect; but one makes progress from the premises to the conclusion; therefore [the conclusion is more perfect than the premises]. I reply: there is nothing absurd in the intellect progressing from more perfect knowledge to less perfect knowledge, with the result that, having grasped both, it itself is rendered more perfect.

[D2.7] *Question [Seven]*[1]: *Must demonstration be made from propositions that are necessary and said of every instance,*[2] *and how?*

[1] Note, first: since necessity is nothing more than a kind of condition bringing unchangeability to things, it is twofold: one is called unqualified,[3] the other natural.[4] The first cannot be impeded by any power, not even by the divine absolute power, because it does not depend on any kind of cause, either intrinsic or extrinsic. By this type of necessity we say that it is necessary for God to exist. The second type cannot be impeded according to the ordinary law of God,[5] but it can be according to his absolute power. By this type of necessity we say that it is necessary [for the sun] to rise and set.[6] Indeed, in the latter type of necessity there are four degrees: to the first are assigned intelligences and rational souls; to the second, the celestial spheres, which depend on intelligences for their movements; to the third, the elements, which are necessary considered in their totality but not in their various parts[7]; and to the fourth, compounds that are dependent on the elements. Here note, moreover, a twofold division in natural necessity: the first is absolute, which is found among things having an intrinsic order to each other, such as man and rational; the other is natural in a qualified way, which is found in things that have an extrinsic order to each other,[8] such as white and swan, etc. Others divide this necessity in a different way, namely, into non-complex and complex[9]: the non-complex pertains to the actual existence of a thing, the complex to a proposition formed by the intellect, and this is the same as the universal posterioristic statement.[10]

[2] Note, second: the posterioristic statement is a proposition in which the predicate goes with the subject and with all things contained under it,

if there are any, always and at any time whatever. In this it differs from the prioristic statement, which is a universal proposition in which the predicate invariably goes with the subject and with everything contained under it, but abstracting from its inherence at all times. They also differ in other ways. First, the names are different, as is obvious: one is said to be the prioristic statement because it looks to the books of the *Prior Analytics* and follows the syllogism according to form in general, and so on; the other is said to be the posterioristic because it looks to the books of the *Posterior Analytics* and follows the syllogism according to matter.[11] Second: the prioristic statement, just as it abstracts from all matter, so also from truth and falsity, and thus it can be found both in the topical syllogism and in the sophistic; the posterioristic, on the other hand, necessarily includes matter and truth. Third: the prioristic statement requires a plurality of things contained under the subject such that an actual distribution of the predicate can be made among the inferiors; the posterioristic, on the other hand, does not require a plurality of subjects, but only that the universal belong with the subject in such a way that it would go with a number of inferiors if there were any; for this reason the proposition "God is unchangeable"[12] is a posterioristic statement.

[3] Note, third: the statement said of every instance, the posterioristic statement, and the commensurate or universal statement[13] are related as higher and lower respectively; for the statement said of every instance is obviously broadest in scope, the posterioristic is less so, and the universal or commensurate lesser still. The proof: every proposition that is essential[14] is also said of every instance, as is apparent in the example, "Every man has sense knowledge." for this proposition is essential, and as a result is said of every instance; whereas the example, "Every man is two-legged" is said of every instance, but it is not essential. The universal or commensurate proposition, on the other hand, is even narrower in scope than the essential proposition, for the same reason; for the proposition "Every man is risible" is commensurate and, as a consequence, essential; whereas the proposition "Every man has sense knowledge" is not commensurate although it is essential. Aristotle makes no mention of the negative posterioristic statement – a proposition in which the predicate is excluded from the subject at any time whatever, exemplified in the proposition "No man is a stone" – because he intended only to treat of ostensive demonstration, which has no use for this negative statement. Also, because this can easily be gathered from what

has been said. Similarly, in the books of the *Prior Analytics* he made no mention of the negative prioristic statement – a proposition in which the predicate is excluded from the subject and all its inferiors, but abstracting from the fact that it does not inhere at all times – for the same reasons.

[4] These matters presupposed, I say: every proper and perfect demonstration of the reasoned fact[15] consists of propositions that are said of every instance and are necessary. Proof of the conclusion: every demonstration of this type makes a thing be known scientifically through its causes by manifesting the properties of its first and adequate subject.[16] But a demonstration cannot do this unless it is made of propositions that are said of every instance, because the property is manifested of its proper subject, and that are also necessary, because to know scientifically is to have knowledge of a thing that cannot be otherwise. Add to this: the causes through which the thing is known are necessary.

[5] You inquire: how is demonstration said to be concerned with a [contingent] effect[17]? I reply: demonstration is concerned with things eternal, not in the sense of things that exist for all eternity, for this is true of nothing except God; but it is concerned with things eternal, that is, with things that have a true connection for all eternity in the divine mind. You object: [a] an eclipse is demonstrated of the moon, and yet an eclipse is not taking place on the moon always and at all times; therefore the conclusion is invalid [and the demonstration is not concerned with things eternal]. Second: [b] there are some predicates that are demonstrated of subjects and yet are not present in them always or are not present at a particular time, as is apparent in things philosophers say of hail and like matters. Some reply to the first objection: the ability to be eclipsed[18] and not the eclipse itself is what is demonstrated. But, to the contrary: mathematicians [demonstrate] eclipses[19] of the moon that are true and real, as is obvious from the middle term they use[20] to prove them.

[6] I reply to the first [5a] and second [5b] objections: said of every instance is threefold.[21] Either it is a proposition in which the predicate inheres in the subject invariably and at all times; or it is one in which the predicate inheres in the subject invariably but not at all times; or it is one in which the predicate inheres in the subject for the most part but neither invariably nor at all times. Demonstrations can be made from all three of these[22]; and from this the solution is obvious.

[D2.8] *[Eighth Question[1]: How many modes of speaking essentially[2]*
are there, and what propositions are contained under them?]

[1] Concerning the first query, I say, first: there are more modes of
speaking essentially than the four Aristotle enumerates in the fifth book
of the *Metaphysics*, text 23.[3] I say, second: the four modes enumerated are
the principal ones, and that is why Aristotle mentioned only them. You
say: why is it, since the first two are the only ones used in demonstration,
that Aristotle was mindful of the others? I reply: because, as he himself
teaches, things that are equivocal should be differentiated into their
principal meanings; and the modes of speaking essentially are equivocal;
therefore [he indicated the four principal meanings]. I say, third: it was
sufficient for Aristotle to enumerate these four modes only, although
others think otherwise; the reason is that three considerations can bear on
any matter: existence, activities, and predicates. If a thing exists in and by
itself,[4] that constitutes the third mode; if it is a cause in and by itself, the
fourth. Predicates are either essential to the thing of which they are
predicated, and so they constitute the first mode; or they are accidental,
and if so, they are either common and rejected as being of little value, or
they are proper, and then they make up the second mode of speaking
essentially.

[2] Concerning the second query, I say, first: propositions contained in
the first mode of speaking essentially are those in which the predicate is
the definition of the subject, or it is the ultimate genus or the ultimate
differentia, or it is a remote genus or a remote differentia.[5] To these can
be added propositions in which the predicate is an extrinsic entity through
which an object is defined, for example, the body in relation to the soul,
for the latter is defined in relation to the former, and a subject with respect
to its properties, for the same reason. Note here, however, that just as
some things are contained primarily in the second mode of speaking
essentially, namely, those in which a property is predicated convertibly
with the subject, as in the examples "Man is risible" and "Animals are
sensible," and others are contained secondarily, e.g., those in which a
property is not predicated convertibly of its subject, as in the examples
"Man is capable of sight" and "Man is a being subsisting *per se*," so some
things are contained primarily in the first mode of speaking essentially,
others secondarily. Contained primarily are those in which a quidditative
predicate is predicated convertibly of its subject, as in the example "Man
is rational"; contained secondarily are those in which a quidditative

predicate is proved non-convertibly of its subject, as in the example "An animal is a living body," etc.

[3] You say, on behalf of Scotus[6]: not only are the foregoing propositions contained in the first mode of speaking essentially but also those in which the genus and the differentia are predicated of the subject, those in which being is predicated of its inferiors, and those in which a subject is predicated of itself, namely, identities. I reply, first: propositions in which being is predicated of its inferiors cannot be contained under the first mode of speaking essentially, first, because propositions in the first mode of speaking essentially must be those in which the predicate is of the essence of the subject; but being is not an intrinsic formality of its inferiors, since it formally implies existence, and this is a thing's modality. Confirmation: because, on the supposition that something is, one may inquire into its essence; therefore existence cannot pertain to the thing's essence. Second: because being can be taken in three different ways: for the essence or quiddity of a thing; for aptitudinal existence; or for actual existence. Taken in the first way it is no different from substance. Taken in the second way, it rather constitutes the second mode of speaking essentially, since aptitudinal existence is a property whose intrinsic formality pertains to the subject. [Taken in the third way] for actual existence – since actual existence is not an accident but rather the formal actuality by which a thing exists – many regard it as reducible to the second mode of speaking essentially, but not to the first. Therefore [existence is not predicated in the first mode of speaking essentially]. I reply, second: identical propositions are not contained under this first mode of speaking essentially, because only propositions whose predicate is of the essence of the subject are placed in the first mode; but in [identical][7] propositions the predicate is not of the essence of the subject. In fact, since there is no more [reason][8] for the predicate to be of the essence of the subject than for the subject to be of the essence of the predicate, they cannot be located even in the second mode. You object, first: Aristotle says in the second book of the *Perihermenias*, last chapter, that the identical proposition "Good is good" is said more immediately[9] than the proposition "Good is not bad"; therefore it is essential. I deny the illation, because by "immediately" Aristotle meant intrinsically and without a middle, not essentially. You object, second: the proposition "Man is man" is not accidental, therefore it is essential. I reply: it is neither essential nor accidental but completely vacuous from the viewpoint of the sciences; and this is gathered from Aristotle in the seventh book of the *Metaphysics*, text 59.[10] So much for that.

[4] I say, second: "proper"[11] can be taken in four ways. It means what belongs to every instance, as man being two-legged; or solely, as man being a philosopher; or to every instance and solely but not at all times, as man becoming gray haired; or to every instance and solely and at all times,[12] as man being risible. Taken in the fourth way it fits the second mode perfectly; in the third way, less perfectly; in the first way, if a property and not a common accident, even less perfectly; in the second way, not at all; and similarly in the first way, if the attribute is a common accident. For this reason all of the following propositions constitute the second mode: "Man is risible," "Man becomes gray haired,"[13] and "Man is capable of sight" – for the attribute is found in every instance, but as a property [whose intrinsic formality pertains to the subject]. On the other hand, the following propositions, "Man has color" and "Man is grammatical," do not constitute the second mode of speaking essentially. Moreover, propositions in which a property is predicated of the definition are placed in the second mode, although some deny this on the ground that propositions placed in the second mode must be those whose subject provides the intrinsic formality of the predicate; but the definition is not the subject of the property; therefore [a property predicated of the definition is not predicated in this way of the subject]. But these [authors] are mistaken: for there are two kinds of subject, one of predication or the metaphysical subject, the other of inherence or the physical subject; and a subject of either type suffices to place a proposition in the second mode. Also located in the second mode of speaking essentially are propositions in which a property that depends on an extrinsic cause[14] is predicated of a subject, as in the example "The moon is eclipsed," for an eclipse depends on the interposition of the earth as an extrinsic cause. The reason for this is that not only is the extrinsic cause placed in the definition of the aforementioned property, but the subject is also[15]; thus, if one wishes to define an eclipse, one must define it as follows: "An eclipse is an absence of light on the moon arising from the earth's coming between it and the sun"; therefore [not only the extrinsic cause, the earth's interposition, but also the subject, the moon whose light is so affected, must be placed in the definition]. You say: when a property is predicated of the extrinsic cause in propositions of this kind, as in the example "An eclipse occurs from the earth's interposition," these too are placed under the second mode. I reply in the affirmative,*16* because the extrinsic cause is of the essence of the property, being placed in its definition.

[5] You ask: are propositions that are convertible in the first mode of

speaking essentially, for example, "Rational animal is man," contained under the second mode? Again, are those in which differentiae are predicated of a genus, for example, "Animal is rational," also included under the second mode? Yet again, are even those in which a property is predicated of its remote subject, for example, "Animal is risible," reducible to the second mode? I reply, first, against Cajetan: propositions of the first kind are not contained under this second mode, because they are indirect and unnatural[17]; only direct and natural propositions are placed under the second mode; therefore [such propositions are not in the second mode]. I reply, second, against him also: propositions of the second kind, whether differentiae are predicated in them under a disjunction or not, are not placed in the second mode of speaking essentially because the genus is neither the proper subject of the differentiae not does it pertain to their intrinsic formality. From this it is apparent *a fortiori* that propositions in which a genus is predicated of differentiae, as in the example "Rational is animal," must be excluded from the second mode also; this likewise for the same reasons, and again because such propositions are indirect and unnatural. I reply, third: propositions of the third kind are not contained under the second mode because those located under the second mode are all posterioristic statements; but propositions in which a property is predicated of a remote subject [are][18] not posterioristic statements, because such properties are not found in everything contained under the subject. For risibility is not found in all animals but only in man.

[6] I say, third: only non-complexes are contained under the third mode of speaking essentially, and not only first substances but second as well.[19] Proof of the first part: for the third mode is one of existing in and by itself, not one of predicating. The second part is against Themistius and Grosseteste,[20] who want to include only first substances under this third mode. They do so because Aristotle said that substances that point out "this thing" are contained under the third mode; and first substances are of this kind whereas second substances are not; therefore [second substances are not contained under the third mode]. Also, substances that are not predicated of another as of a subject are contained under this {third} mode[21]; but second substances are predicated of first substances as of a subject; therefore [they are not located in the third mode]. But, to the contrary: Aristotle said[22] that substances without qualification are contained under this third mode, and even those that point out "this thing." Again, Aristotle excluded from this third mode only those things

that are predicated of a subject of inherence, and accidents are of this kind whereas second substances are not.

[7] I say, fourth and last: all four causes[23] are contained under the fourth mode, because the definition of a cause given by Aristotle is applicable to each of the causes; therefore [each cause is contained under the fourth mode]. Proof of the antecedent: for Aristotle's definition is the following: "A cause is that through which one thing is present in another"; therefore [the definition is applicable to all four causes]. Again, causes involve an intrinsic relationship to what they cause; but, as such, they can only be contained under the fourth mode; therefore [all four causes are contained under the fourth mode]. Proof of the minor premise: they are not contained in the third mode, as is obvious; nor in the first and the second, because these modes are essentially constituted by an intrinsic relationship that involves existing and predicating; [therefore they must be contained under the fourth mode].

[8] You ask, first[24]: is a cause put in the fourth mode with respect to an effect in potency as well as with respect to an effect in act? I reply: it is put there with respect to an effect in potency, because an effect in potency, precisely as involving a relationship to its cause, is not included in the first mode or in the second, therefore in the fourth. Again, it is contained there with respect to an effect in act, because Aristotle teaches this in the example he gives of throat-cutting.[25] You say: a cause as it respects an effect in act is contingent,[26] therefore it cannot be placed in the fourth mode. I reply: such a proposition is indeed contingent if one focuses on the predication; it is not contingent, however, and this is our concern, if one focuses on the connection of the predicate with the subject and the causal relationship between the two.

[9] You ask, second: is the fourth mode of causing also one of predicating[27]? I reply, first: there is nothing to prevent one and the same proposition from being in the first and the second mode, and also in the fourth, when taken in a different way. This can be seen in the proposition "A rational animal is discursive," for this, when considered from the viewpoint of the subject's being of the essence of its predicate, is in the second mode of speaking essentially; yet viewed from the perspective of the relationships between cause and effect, it is in the fourth mode. I reply, second, with common opinion and against Apollinaris[28]: the fourth mode of causing is not one of predicating. This is so because Aristotle, having enumerated the first and second modes of speaking essentially, teaches that things predicated of another subject are predicated

accidentally; but effects are predicated of their causes as of another subject; therefore, they cannot constitute the first and second modes, which are essential. Again, because it would follow that all propositions of the fourth mode would be essential, and this is false, for many of them are contingent. For the fourth mode differs from the first and the second in this, that the first and the second mode require a subject that is particular to them and determinate, either as a genus or as a species, as is apparent in the proposition "Man is a rational animal, and risible," and so on; the fourth mode, on the other hand, does not require such a proper subject, and for this reason the proposition "A hot thing heats" is in the fourth mode of speaking essentially, being understood of anything hot whatsoever that heats either essentially or accidentally.

[10] You ask, third: why did Aristotle not enumerate four modes of causing according to the four species of cause just as he did the two modes of predicating essentially? I reply: because it is quite easy for anyone to recognize the four modes of causing, but not the two of predicating.[29] Second: the two modes {of predicating}[30] serve a useful purpose in demonstration, whereas the four modes of causing do not; therefore [he enumerated only those of predicating].

[D2.9] *Ninth Question[1]: What are the rules for recognizing propositions contained in the first and second modes,[2] and are there more than two modes of predicating[3]?*

[1] Concerning the first query, I reply: the following are the principal rules. The first: propositions of the first and second modes must be necessary by at least a natural necessity.[4] From this it follows that the propositions "Peter is rational, risible, etc." are neither in the first nor in the second mode, because they concern a contingent subject.[5] But note: if the foregoing propositions are considered with respect to the connection of the predicate with the subject, such that the word "is"[6] indicates not existence but the connection of the subject with the predicate, they are in the first or in the second mode depending on whether the predicate pertains to the essence of the subject or vice versa.

[2] Second rule: propositions of the first and second mode must be statements said of every instance[7]; from this it follows that the propositions "Animal is risible, wondering, etc." are not of the first and second mode.

[3] Third rule: propositions of this type must be such that the predicate

is said of the subject directly and naturally. Hence the proposition "Animal is man" would not be in the first or second mode; nor would one in which an accident is defined in the concrete through its order to a subject, such as the proposition "Risible is man capable of laughing." However, these propositions are reducible[8] to those of the first and second modes.

[4] Fourth rule: in propositions of the first and second mode the predicate must pertain to the definition of the subject, or vice versa. From this it follows that many propositions common opinion would place in the first or in the second mode are actually not contained under them, as is apparent in the proposition, "A body is in place essentially."[9] This is not contained in either the first or the second mode, and nonetheless the predicate is in the subject immediately and necessarily.

[5] Fifth and last rule: in propositions of this type the predicate must contain a true and perfect definition of the subject,[10] or vice versa. For this reason the propositions "A vacuum is a surface devoid of bodies" and "A chimera is a kind of impossible being" are not in the first or second mode, because the predicate does not contain a true and perfect definition of the subject, or vice versa.

[6] You object: in the second book of *De caelo* Aristotle proves[11] that the heavens are spherical and that the earth is round, and yet roundness is not an intrinsic formality of either the heavens or the earth; therefore our teaching on what constitutes propositions of the first and second modes that serve the purpose of demonstration is not correct. I reply: the demonstrations Aristotle offers in the second book of *De caelo* concerning the roundness of the heavens and the earth are not true and perfect demonstrations,[12] because they manifest a property of the subject that is improper[13] and not proper.

[7] Concerning the second query: there seem to be more than two modes of predicating essentially[14] because [a] there are five predicables,[15] and thus there ought to be five modes of predicating essentially. Again, [b] the mode of predicating accidentally[16] is threefold: for either an accident is predicated of its subject, as "Man is white"; or vice versa, as "White is the swan"; or one accident is predicated of another, as "White is sweet"; therefore there ought to be three modes of predicating essentially. Proof of the consequent: because the modes of predicating accidentally are opposite those of predicating essentially; but the same number of modes attributed to one of a pair of opposites should be attributed to the other; therefore [there should be three modes of

predicating essentially]. Third: [c] at least some propositions of the fourth mode are not of the first or the second mode; therefore [there are more than two modes of predicating essentially.]

[8] Our conclusion: the modes of predicating essentially enumerated by Aristotle are only two in number. The proof: from the authority of Aristotle, who teaches this; and from reason, because a mode of predicating is taken from a mode of being,[17] and the modes of being are twofold, for either the predicate pertains to the essence of the subject, or vice versa; therefore [there are only two modes of predicating essentially].

[9] To the first objection [7a], I deny the inference: because the predicables are differentiated not on the basis of modes of predicating essentially but rather on the modes of predicating either in quiddity,[18] or in quiddity of a qualitative kind,[19] or in quality convertibly,[20] or in quality non-convertibly[21]; therefore [the modes of differentiating predicables are not the same as those of differentiating essential predications].

[10] To the second [7b], I reply: there should be as many modes attributed to one pair of common opposites as there are to the other; thus it happens that just as the modes of speaking essentially are threefold, namely, either in predicating or in existing or in causing, so also the modes of speaking accidentally should be threefold, as we have shown. But it is not true that the same number of modes as are said of one particular member contained under a common kind of contrariety need be said of the other. Thus the argument is not valid.

[11] To the third [7c], I reply: propositions of the fourth mode are not formally those of predicating, but of causing.[22] Thus if any are said to be of the first and second mode, this is because their predicate pertains to the essence of the subject, or vice versa.

[D2.10] *Tenth Question[1]: What are the modes[2] that serve the purposes of demonstration[3]?*

[1] Albert[4] says that all four modes serve the purposes of demonstration; St. Thomas and Cajetan reject the third mode; Philoponus, Themistius, and Averroes reject the third and the fourth.

[2] Note[5] that a proposition of a demonstration can be considered in three ways: either with respect to the terms of which it is composed; or with respect to the cause it involves; or with respect to predication.

[3] I say, first: the first and second modes of speaking essentially serve the purposes of all perfect demonstrations, because Aristotle teaches this

in no uncertain terms; also because a perfect demonstration must be composed of propositions that make use of essential predication,[6] and propositions of the first and second modes are of this type; therefore [both these modes serve the purposes of perfect demonstrations].

[4] I say, second: the third mode, the mode of existing in and by itself,[7] does not serve the purposes of demonstration in the proper sense, for a demonstration is composed of propositions[8] whereas this mode applies properly to substances; it can serve such purposes in an improper way, however, that is, by reason of the terms of which a particular proposition in the demonstration is composed, for these can sometimes designate substances.

[5] I say, third: the fourth mode enters into demonstration. Proof, first: every perfect demonstration of the reasoned fact must be composed of propositions in which the true cause of a property is contained; therefore [such a demonstration requires the fourth mode, the mode of causality]. Second: the middle term in a most powerful demonstration is either the definition of the subject or the definition of the property; if the first, since the definition is related to the thing defined as a formal cause, which pertains to the fourth mode, it follows [that a most powerful demonstration of the first kind requires the fourth mode]; if the second, since the subject or the definition of the subject is the efficient cause[9] of the property, and the efficient cause pertains to the fourth mode, the same result follows [namely, a most powerful demonstration of the second kind requires the fourth mode].

[6] Objection: propositions of the fourth mode include only extrinsic causes[10]; therefore they cannot serve the purposes of perfect demonstration, which contain only intrinsic causes. I reply: it is false to say that propositions of the fourth mode contain only extrinsic causes, since they also contain intrinsic causes.[11] Second, I reply: a perfect demonstration can sometimes be composed of extrinsic causes. Nor can you say: propositions of the fourth mode containing intrinsic causes are either in the first or in the second mode of speaking essentially, for, seeing that the first and second modes are modes of being and of predicating, they cannot formally be modes of causality, which is found only in the fourth mode.   ·

[7] Second objection: Aristotle teaches in text 10[12] that perfect demonstration contains only two modes of speaking essentially. Confirmation: the same text teaches that demonstrations must be composed of modes of predicating essentially, and these are only two in

number. I reply: Aristotle meant there that demonstration takes its necessity from the first and second modes of speaking essentially. This is clear, because he had said at the beginning of the chapter[13] that demonstration must be composed of necessary propositions, and in text 9[14] he enumerated four modes of speaking essentially. Therefore all that remained was for him to show what modes of speaking essentially would give a demonstration its necessity. To the confirmation, I reply: in that text Aristotle teaches that demonstration must be made only from the first and the second mode of speaking essentially because demonstration must be made from essential propositions, and these alone are those in either the first or the second mode. You ask: if this is so, why did Aristotle not mention the fourth mode? I reply: because not all propositions of the fourth mode serve the purposes of perfect demonstration; again, because one could gather from text 5[15] that the fourth mode does enter into demonstration, because there Aristotle had taught that demonstration must be made from causes.

[D2.11] *Eleventh Question: What is a universal predication,*[1] *and what propositions are contained under it?*

[1] Concerning the first query, Aristotle defines universal predication as that which is said of every instance, essential, and commensurate.[2] To understand this, note, first: essential is less universal than said of every instance, and commensurate is still less univeral than said of every instance, as we have explained above.[3] For this reason, when Aristotle teaches in text 11[4] that the proposition "A line is straight" is commensurate but that it is not said of every instance, some other property must be implicitly understood, such as "or curved." Thus the meaning would be that the proposition "A line is straight or curved" is commensurate or universal.

[2] Note, second: universal is taken here not for a nature that is apt to exist in many or to be predicated of many, the way it was taken by Porphyry, but for what belongs with an adequate subject[5] and does so convertibly.

[3] Note, third: universal or commensurate[6] is what belongs to a thing precisely as such,[7] the way, for example, "rational" occurs in the proposition "Man is rational." Proof: from Aristotle, in text 11,[8] where he teaches that the universal is manifested of its primary and adequate subject, and in text 14,[9] that the universal is convertible with its subject.

These statements would not be true if what we have maintained were not true also. Hence one may see the error of those who hold that a predicate is universal or commensurate if it belongs to a subject by virtue of one of its essential parts, the way, for example, "sensible" occurs in the proposition "Man is sensible." You object: if this is so, why does Aristotle teach in texts 11 and 19[10] that essential propositions are universal and commensurate? I reply: in those texts Aristotle means only that essential propositions in a proper and perfect demonstration are universal and commensurate, since elsewhere he likewise defines universal, saying that it is what is said of a primary subject[11] and, through it, of other things also. This definition, like the preceding, is obvious from the foregoing.

[4] Concerning the second query, many authors have many opinions.[12] Some understand universal propositions or those placed under a universal predicate to be immediate. Others hold that they are those in which all proper attributes are predicated of their subjects. Themistius maintains that they are those in which the predicates belong to the subject with no intermediate cause coming in between. So, according to him, the proposition "Man is an animal" has a commensurate predicate, whereas the proposition "Man is risible" has not, because there is an intermediate cause between man and risible, namely, rational.

[5] Note: there are many and various grades[13] for each and every thing. Some are common, for example, with respect to man: to be a substance, to be a living body, to be an animal, and so on. Others are proper, and through these objects are constituted in their being and set apart from all others; such, for example, is the grade of rationality with respect to man. Many proper attributes follow from this proper grade, for consequent on rationality are discourse, wonder, and so on.[14]

[6] Note, second: it is one thing to belong to an object essentially, another to belong to it precisely as such[15]; for common grades belong to an object essentially and they are predicated of it in the first mode,[16] but they do not belong to the object precisely as such.

[7] These matters understood, we say: universal propositions, or those that are commensurate,[17] are those in which the predicates belong to the object precisely as such[18] or according to its proper formality; so hold Averroes, Philoponus, and Albert. The proof: from Aristotle, in texts 11 and 14,[19] where he states that propositions in which the predicate is said convertibly of the subject are universal, such that, if the predicate were removed, the primary subject would be removed also. You ask: why are such predicates said to be universal and primary? I reply: because they are

found in everything contained under the subject and in all its universality, as is apparent in the proposition "Man is rational"; and because they also belong with their subjects adequately and convertibly.

[D2.12] *Last Question*[1]*: Must perfect demonstrations*[2] *be made from propositions that are essential, universal, and proper*[3]*?*

[1] It seems that demonstration is not made from essential propositions: [a] because Aristotle treats of chance and fortune in the second book of the *Physics*, and these are accidental kinds of being.[4] Again, [b] in the *De caelo* Aristotle manifests many accidents of subjects that are not proper accidents, for example, that the heavens move circularly, that the earth is at rest, that fire is the primary hot body, air the primary moist body, and so on. Yet again, [c] there are many accidents that do not have a proper cause and yet are demonstrable through a cause from which they derive their necessity.

[2] Second: [a] it seems that demonstration is not made from universals,[5] because not every accident has a proper subject with which it is convertible. Again, [b] otherwise it would follow that demonstrations composed of essential propositions would not be true demonstrations, against the common opinion. For the demonstration, "Every animal is sensitive; every man is an animal; therefore, [every man is sensitive]," is a true demonstration, and yet it is not composed of universals.

[3] Alfarabi, cited by Averroes in the first book of the *Posterior Analytics*, teaches that demonstrations can sometimes be made from proper propositions that are not primary,[6] sometimes from primary propositions that are not proper, and sometimes from propositions that are neither primary nor proper, as is that in which we prove that the earth is round. Averroes says that every demonstration must be made from essential and universal propositions. The proof: [a] from the authority of Aristotle when discussing demonstration in general; again, [b] demonstration follows nature, whence, just as a property is present naturally in its primary and adequate subject, so a demonstration ought to demonstrate that property from its primary and adequate subject; and yet again, [c] otherwise it would follow that demonstration could be composed {of} propositions that are merely accidental.[7]

[4] I say, first: every demonstration of the reasoned fact must be composed of essential propositions, because Aristotle clearly teaches this in chapters 4 and 6. Also, demonstration of the reasoned fact must

proceed from a true and proper cause[8]; but it cannot proceed from a proper cause if it is not composed of essential propositions; therefore [it must be so composed]. The minor premise is proved by induction.[9]

[5] To the first objection to the contrary [1a], I reply: an accidental kind of being can be considered in two ways: either as it is merely an accident, and as such it is eliminated from all the sciences; or for the reason that properties of a certain type are associated with it, such as being beyond the intention of the agent, or happening rarely.[10] In this second way it can fall under a science. Aristotle treats of chance and fortune in the second way in the second book of the *Physics*.

[6] To the second objection [1b], I reply: natural circular motion is a proper attribute of the heavens; rest, on the other hand, is not proper to earth, and thus the demonstration by which Aristotle proves that the earth is at rest is not a true demonstration[11]; moreover, being the primary hot body and being the primary moist body may be considered as the most proper attributes of fire and air.

[7] To the third objection [1c], I reply: if accidents[12] did not have a proper cause they could not be demonstrated in a proper sense, since a proper cause is required for a perfect demonstration.

[8] I say, second: a most powerful demonstration must be made from propositions that are essential and universal,[13] for this is what Aristotle teaches in text 11[14] and in chapter 6. Also, because otherwise it would follow that one could construct a number of most powerful demonstrations while keeping one and the same middle term, which is quite absurd. Proof of the inference: if one were to use the definition of animal to demonstrate a particular property, by keeping the same middle term one could apply the same demonstration to all living species; and so there would be as many demonstrations as there are species of animals. This is apparent in the following demonstration:[15] "Every corporeal sensitive nature is apt to sense and to move in all different directions; but every animal is of this type; therefore [every species of animal is apt to sense and to move in all different directions]." Again, every most powerful demonstration is indeed most perfect, and so there cannot be a demonstration more perfect than it is; but a demonstration that would be made of propositions that are essential and universal would be more perfect than one that is made only from propositions that are essential; therefore [a most powerful demonstration must be composed of propositions that are both essential and universal]. From this it follows that both premises of a demonstration must be universal and necessary,

and this on the part of the subject and on the part of the predicate.[16] From this it results that mathematical demonstrations, which for the most part have one premise that is common, are not perfect.[17] And if they are said to be perfect, this is because of their highest degree of certitude, since they abstract from matter, which [is] the cause of uncertainty[18]; or else it is because of the preeminent method they follow.

[9] Note here: apart from having both premises universal propositions, most powerful demonstration also requires that both premises be proper. This is obvious, partly from the foregoing, partly because most powerful demonstration proceeds from the definition[19] of the subject or from the definition of the property; but it cannot proceed from these unless it is made from proper premises; therefore [most powerful demonstration also requires proper propositions in its premises].

[10] I say, third and last: demonstration that is true and proper but not most powerful can be made from propositions that are not universal[20] provided they are essential. Proof of the conclusion: in a demonstration an attribute that belongs to an object by reason of some higher genus can sometimes be demonstrated, as is apparent in the demonstration, "Every animal is sensitive; every man is an animal; therefore [every man is sensitive]." Here sensitivity is demonstrated of man, but this belongs to man not insofar as he is man but insofar as he is an animal, which pertains to a higher genus; therefore [a demonstration can be made from propositions that are essential only]. Yet all are agreed that this is a true demonstration.

[11] To Averroes's first objection [3a], I reply: Aristotle is speaking of most powerful demonstration.

[12] To the second [3b], I reply: this applies to most powerful demonstrations, and we agree on this also.

[13] To the third [3c]: I deny that propositions that are not universal are merely accidental.

[14] To the two objections opposed to the contrary [2a,b] I reply: proving the third conclusion does not militate against the second conclusion.

[15] From these conclusions it follows that there can be neither demonstration nor science of individuals,[21] because they are uncertain, indeterminate, changeable, and inclusive of many extrinsic accidents, whereas demonstration must consist of universal and unchangeable propositions, either negatively or by reason of the connection of the predicate with the subject, which has truth from all eternity in the divine

mind. But there can be demonstration and science of God in the way in which these can be in us; because God is certain, determinate, unchangeable, and devoid of all accidents. And if God is singular,[22] his very singularity pertains to his intrinsic definition.

[D3] *Third Disputation: On the Species of Demonstration*[1]

[D3.1] *First Question: How many species of demonstration are there?*

[1] Note, first: demonstration can be taken in two ways: for a syllogism established from necessary propositions, proper or common; or for a syllogism constituted from necessary and proper propositions only, in the way it is taken by Aristotle in the first book of the *Posterior Analytics*. Demonstration taken in the first way, when compared with a proper and perfect demonstration or with a syllogism taken in the second way, is said to be topical reasoning by Aristotle[2]; when compared with probable reasoning, on the other hand, it is said to be demonstration from the fact that it proceeds from necessary propositions.

[2] Note, second: here we are not inquiring into the species of demonstration as it is a syllogism. In the latter way it can be divided by reason of quantity into universal and particular, by reason of quality into affirmative and negative, and by reason of mode of inference into ostensive and reducing to the impossible. Here we are seeking species of demonstration taken in the second way[3] and insofar as it is an instrument of scientific knowing.

[3] Note, third: the species of demonstration that makes us know the cause of any effect as it is from its very nature also makes us know the existence of the effect, for we cannot know the cause of any effect unless we know the existence of that effect at the same time. From this it is apparent that Averroes[4] divides the species of demonstration erroneously on the basis that one species makes us know the cause of an effect, another the cause and existence of the effect.

[4] It is certain, first, that demonstration to the impossible is not true and perfect demonstration, because it proceeds from false premises and denies both premises in the process of questioning. It is certain, second, contrary to certain moderns whose mentor was Franciscus [Neritonensis],[5] that demonstration of the reasoned fact is true and certain demonstration, for otherwise Aristotle would have labored in vain when assigning its properties, and because demonstration of the reasoned fact generates true science.

[5] The first opinion is that of Avicenna, cited by Averroes in many places,[6] holding that there is only one species of demonstration, namely, demonstration of the reasoned fact. Proof, first, on the authority of Aristotle: [a] in the second book of the *Prior Analytics*[7] Aristotle states that when we prove the prior through the posterior we beg the question[8]; but in demonstration of the fact we prove the prior through the posterior; therefore [demonstration of the fact begs the question and is not a valid species]. Second: [b] in this book, chapter 2, Aristotle teaches that we know in an unqualified way and not sophistically when we know an object through the cause that makes it exist; but we do not know an object through the cause that makes it exist through demonstration of the fact; therefore demonstration of the fact is not true demonstration. Third: [c] in the same place Aristotle lists all the properties of a true and perfect demonstration, and after enumerating them he says that without them there will be a syllogism but not a demonstration; but these properties do not square with those of demonstration of the fact; therefore [demonstration of the fact is not true demonstration]. Fourth: [d] Aristotle teaches in the third chapter that there is no circular demonstrative regress,[9] for otherwise there would be two ways to know scientifically, and such knowing would be defined inadequately; but if demonstration of the fact be conceded, there would be a circular demonstrative regress and a twofold scientific knowing, and it would follow that scientific knowing would be defined inadequately; therefore [demonstration of the fact is not true demonstration]. Confirmation: [e] in the same place Aristotle teaches that demonstration must proceed from things more known with respect to nature; but demonstration of the fact proceeds from things more known only with respect to us[10]; therefore [demonstration of the fact is not true demonstration].

[6] Proof, second, from reason: first, [a] demonstration must be established from direct and natural propositions; but demonstration of the fact is made from propositions that are indirect and unnatural[11]; therefore [demonstration of the fact is not true demonstration]. Proof of the minor premise: from the example of the demonstration "Every risible [being] is a rational animal[12]; but every man is risible; therefore [every man is a rational animal]." Second: [b] if demonstration of the fact were a species of perfect demonstration, something would be proved in it; but nothing is proved in it and it begs the question[13]; therefore [demonstration of the fact is not a species of perfect demonstration]. Proof of the minor premise: I take the foregoing demonstration and argue as follows. Either

I know in the premises that risibility is necessarily connected with rationality or I do not. If I do not, then, since the demonstration of the fact does not proceed from necessary premises, it will not be a demonstration. If I do know it, since the connection cannot be known without knowing its cause, before having proved risibility to exist in man one must already have known this scientifically, and that is to beg the initial question. You say: we know that risibility exists in a rational animal by sense knowledge and by induction. To the contrary: induction does not prove anything necessarily, and an essential proposition such as "A risible [being] is a rational animal" must be known essentially and not through the senses.[14] You say, second, with Averroes: risibility is known to exist in a rational animal because it is placed in the definition of the subject. But, to the contrary: for risibility cannot be known to be placed in the definition of the subject without its cause and its existence also being known; therefore such a demonstration necessarily begs the question. And this is the first opinion. A final confirming argument: [c] in the major premise the cause is known from its effect; therefore it is pointless to inquire about its existence.

[7] The second opinion is that of Averroes in the first book of the *Posterior Analytics*, in the *Epitome of Logic* in the chapter on demonstration, and in the first book of the *Physics* in the prologue, and elsewhere; of Zimara in his entry on demonstration and in his *Theoremata*; of Balduinus in his *Quaesita*; and of others[15] maintaining that there are three species of demonstration: of the fact, of the reasoned fact, and most powerful. Demonstration of the fact proves a cause from its effect; demonstration of the reasoned fact proves only the cause of the effect and not its existence, since this is know either through the senses or in some other way; and most powerful demonstration proves both the cause and the existence of the effect.

[8] Proof of this opinion, first, from Aristotle: [a] from texts 5 to 30[16] he treats of most powerful demonstration only; then, from text 30 onwards[17] he treats of demonstration of the fact and demonstration of the reasoned fact; therefore [there are three species of demonstration]. Second, [b] in text 42[18] he teaches that a science that makes us know the cause and existence of something is the better science; but such a science is obtained through most powerful demonstration; therefore [most powerful demonstration is a species apart from the other two]. Third, [c] in the second book of the *Posterior Analytics*, text 7,[19] Aristotle teaches that there is a middle term through which we can know the cause and the

existence of something, and this can only be found in most powerful demonstration. Again, from reason, first: [d] the enumeration of species of demonstrations should be taken from the enumeration of middle terms on which a demonstration intrinsically depends; but there are three types of middle term that are distinct one from the other; therefore there are three species of demonstration. Proof of the major premise: in the first book of the *Topics*, chapter 1,[20] Aristotle argues to three species of syllogism from a threefold distinction of principles; therefore, for the same reason, it is legitimate to argue to three species of demonstration from a threefold distinction of principles. Proof of the minor premise: the middle term of a demonstration is more known only with respect to us or "in knowing"; or it is more known only with respect to nature or "in being,"[21] and of this kind are physical causes; or it is more known both with respect to us and with respect to nature, or "in knowing" and "in being"; therefore [there are three types of middle term distinct one from the other]. Second, [e] three things are being sought, therefore three species of demonstration. Proof of the inference: the enumeration of instruments is to be taken from their end; but the knowledge of what is being sought is the end of demonstration; therefore, if three things are being sought, there are three species of demonstration. Proof of the antecedent: either the existence of something alone can be sought, or the cause of something alone, or both together. Nor can you object: even more kinds of knowledge are being sought, namely, the quiddity of the thing and its qualities, and so, if the kinds of demonstration are to be based on the enumeration of things being sought, there should be five species of demonstration just as there are five knowledges being sought. But this is contrary to the opinion of all: because the quiddity of something cannot be proved through demonstration, for the quiddity has no cause through which it can be proved; and its qualities are reducible to a complex existent[22]; therefore [neither the quiddity nor its qualities constitutes a special kind of knowledge being sought].

[9] The third and last opinion is that of Themistius on the first book of the *Posterior Analytics*, chapter 27; of Philoponus, on text 30; of Algazel in his logic; and of Thomas, etc.,[23] who hold that there are only two species of demonstration, of the fact and of the reasoned fact. First, because Aristotle teaches this expressly in text 30[24] of this book; second, because we know something either *a posteriori* or *a priori*, *a posteriori* through demonstration of the fact, *a priori* through demonstration of the reasoned fact.

[10] I say, first: demonstration of the fact is a true species of demonstration. The proof: from the authority of those cited, to whom can be added Alexander[25] on the first book of the *Topics*, chapter 1, who, while teaching that demonstration of the fact is called a topical syllogism when compared with perfect demonstration, nonetheless holds that it is true demonstration. Also, from the authority of Aristotle, as above. Again, because demonstration of the fact proceeds from necessary premises and infers a necessary result; and it does not generate opinion; therefore, it generates science. You object: Aristotle teaches that demonstration must be made from universal premises, and so on; but if demonstration of the fact, which does not fulfill these conditions, were to be a true species of demonstration because it infers a necessary result from necessary premises, then induction also, since it infers a necessary result and does not generate error or opinion, would be true demonstration. I reply: demonstration of the fact is a true species of demonstration even though it is less perfect than demonstration of the reasoned fact; but induction is not, because is not made from propositions that serve the needs of science,[26] the way in which demonstration functions as an instrument, since it argues from singulars. Add to this: induction of and by itself concludes nothing necessarily.

[11] I say, second: Averroes's most powerful demonstration is not a species distinct from demonstration of the reasoned fact. First, because demonstration of the reasoned fact, by the very way it works,[27] makes us know the cause and the existence of something; therefore Averroes's most powerful demonstration, making us know nothing in addition to the cause and the existence of something, is not a distinct species. Confirmation: the middle term of a demonstration is twofold, and from this one should take its differentiation; for either it is "of knowing" only, and this constitutes demonstration of the fact, or it is additionally "of being"; but since a thing is related to being as it is to knowing,[28] if the middle term is a cause of being it will also be a cause of knowing; therefore [there are only two kinds of middle term, "of knowing" and "of being and of knowing," and so there are only two species of demonstration]. Again, otherwise it would follow that one and the same demonstration would differ from itself in species, which is absurd.

[12] A second proof: because one and the same demonstration would be a demonstration of the reasoned fact for someone who knew the thing's existence and most powerful demonstration for another who did not. Add to this: just as demonstration must be made from premises that

preferably are more known with respect to nature, as we have taught, so also must definition, which is its principle, as Aristotle teaches in the sixth book of the *Topics*.[29]

[13] I say, third: there are only two species of demonstration, of the fact and of the reasoned fact. This follows from the foregoing, for demonstration of the fact is true demonstration, as we have shown, and most powerful demonstration is no different from demonstration of the reasoned fact; therefore [there are only two species]. Confirmation: for we seek either something's existence through demonstration of the fact, or its cause and its existence through that of the reasoned fact.

[14] I say, fourth: demonstration of the reasoned fact contains under it two subspecies of demonstration, one proceeding from extrinsic causes, the other from intrinsic causes and manifesting a property of its primary and adequate subject through principles that are actually indemonstrable; and this second type can, with full justice, be called most powerful. Proof of the first part of the conclusion: demonstration in general is made to be what it is from the fact that it proceeds *a priori* from true causes; but there are two kinds of cause, intrinsic or extrinsic[30]; therefore [there are two kinds of demonstration]. Proof of the second part: a demonstration that fulfills these conditions is the most perfect of all, and so it can with full justice be called most powerful. You object against these two conclusions: in the first book of the *Posterior Analytics*[31] Aristotle calls demonstration to the impossible true demonstration and compares it with demonstration of the reasoned fact; therefore there are more species still. I reply: Aristotle calls this demonstration, but demonstration to the impossible, and he compares it with demonstration of the reasoned fact to show that it is different in the greatest way possible; therefore [demonstration to the impossible is not another species of ostensive demonstration]. You object, second: demonstration of the reasoned fact is vacuous, because for it one must know the cause of its effect and therefore one cannot prove it. I reply: one must know the cause either in the premises or, prior to the premises,[32] one must know its necessary connection with the effect so that in the demonstration one can give the reason why it is present in the subject.

[15] To the objections from the first opinion: to the first [5a], I reply: we beg the question[33] when we prove the prior through the posterior either with respect to nature or with respect to us; or, second, when we beg the question with respect to nature only, not with respect to us. To the second and the third [5b, c], I reply: Aristotle is there discussing perfect science

and demonstration, and if demonstration of the fact is compared with this it will be a topical or probable syllogism[34]; if, on the other hand, it is considered in itself it is true demonstration because it infers a necessary result from necessary premises and does not generate error or opinion. To the fourth [5d], I reply: Aristotle is aware in that text that he is arguing against the ancients,[35] who proposed that we attain perfect science by arguing demonstratively in a circle[36]; for this reason he insists correctly that such a circle is not possible, because scientific knowing would be defined improperly, since it would be twofold. Supply "perfect" [when defining scientific knowing] and one begs the question.[37] To the confirmation [5e]: the reply is obvious from the foregoing.

[16] To the first argument [6a], I reply: demonstration of the fact, precisely as imperfect, can be constructed of propositions that are indirect and unnatural.

[17] To the second [6b], I reply first, indirectly: if the argument were valid it would also remove the possibility of having demonstration of the reasoned fact, because in that type of demonstration one must also foreknow the connection between cause and effect, and for this reason the same argument [would apply to it]. I reply, second and directly: we know the connection of a property with its subject by experience, for, from the foundation of the world to our own times risibility has always been found with man; second, by induction, for it is true to affirm of each and every man that he is risible; third, by the light of the intellect,[38] which recognizes that this connection is necessary, for whatever happens for the most part and always is natural, and, since risibility is always present in man, the intellect recognizes this as natural. And it can be confirmed from this, for otherwise nature would have looked unkindly on its universality by not having given things their proper conditions and necessary properties.[39] And from this the reply to the second argument is apparent. To the confirmation [6c], I reply: the cause is not known from the effect, but rather the connection of the cause with the effect is known in the major premise.

[18] To the first argument of the second opinion [7a], I reply: from text 5 to 30[40] Aristotle deals with demonstration of the reasoned fact, though some things he says there should be understood [of] most powerful demonstration,[41] which we concede. To the second [7b], I reply: the text of Aristotle is not against our position, because a science that has knowledge of existence only is less perfect than one that has knowledge of cause and existence, and the latter is what is had through demonstration

of the reasoned fact. To that particular text I reply: this is not to the point, for Aristotle teaches in the second book of the *Posterior Analytics*[42] that there are four kinds of cause, some of which are intrinsic, some extrinsic, some convertible, others not; and true demonstration can be made from all of them. To the two arguments [7d, e], the reply is obvious from the foregoing.

[D3.2] *Second Question: How are demonstrations of the reasoned fact and demonstrations of the fact similar and dissimilar,*[1] *and on the division of the latter.*[2]

[1] Concerning the first query, I say: demonstration of the reasoned fact and demonstration of the fact are analogically the same,[3] because the latter and the former both proceed from true and necessary premises, and because demonstration of the fact has much the same properties as demonstration of the reasoned fact. If, however, they are expressly considered from the viewpoint of their being made from convertible terms, as Averroes teaches in the *Epitome of Logic* and in the first book of the *Posterior Analytics*, they differ by reason of their middle term and by reason of their end, as is obvious in itself; and as a consequence they differ essentially, whatever others may say. This is proved from Aristotle, who explains the differentiae between demonstration of the reasoned fact and demonstration of the fact in text 30[4] of this first book; but differentiae are found only between things that are distinct in species. And, although one and the same numerical conclusion can be demonstrated through demonstration of the reasoned fact and through demonstration of the fact, it cannot be so demonstrated formally; because a conclusion has an intrinsic ordering to the middle term through which it is proved; but the middle term of a demonstration of the reasoned fact, being a middle "in being" and also "in knowing," is specifically distinct from the middle term of a demonstration of the fact, which is a middle "in knowing" only; therefore [a conclusion reached by the one is not formally the same as the same conclusion reached by the other].

[2] Hence it comes about that demonstration of the reasoned fact usually is defined differently from demonstration of the fact, and the two are called by different names. For, considered from the viewpoint of its middle term, Aristotle calls demonstration of the fact demonstration "of a sign"[5] in text 19[6] of this book; considered from the viewpoint of its end, he calls it demonstration "of the fact" or "by which,"[7] since it proves the

existence of something. Considered from the viewpoint of its mode of proceeding, Averroes calls it demonstration "of evidence,"[8] since it proceeds from what is more known with respect to us; he also refers to it as demonstration "of existence."[9] The Latins call it demonstration "from an effect" or "*a posteriori*"[10]; the Greeks call it "conjectural"[11] demonstration.

[3] Concerning the second point, I say: there are many divisions of demonstration of the fact. A first division: one kind proceeds from remote causes, and this is not of concern here; another proceeds from effect to cause. An example of the first: a wolf does not reason,[12] therefore it has no sense of wonder; an example of the second: there is smoke, therefore fire.[13] But note here: this second kind of demonstration can proceed from effect to cause, or from one effect to another, or from a sign or from any accident necessarily connected with the cause to that cause. A second division is taken from the middle term: one kind is made from convertible terms, for example, an eclipse occurs, therefore there is an interposition of the earth; another kind from non-convertible terms, for example, heating occurs, therefore there is fire.[14] A third division: one manifests a simple existence, the kind in which Aristotle proves the existence of protomatter,[15] a first mover,[16] and [the element] fire[17]; another manifests a complex existence, the kind that show a certain proposition to be true *a posteriori*, such as those that prove man has a sense of wonder or is rational. But note: demonstrations of this kind are most useful in the sciences,[18] because the principles of such sciences are sometimes not known and they cannot be proved except through demonstrations of this type. Second: because without their help we would know practically nothing about abstract and divine matters. Add to this: they are more familiar to us.

[D3.3] *Third Question: Is there a demonstrative regress[1]?*

[1] The first opinion is that of ancient philosophers, referred to by Aristotle,[2] who thought there should be a perfect circle[3] in demonstration in such a way that the conclusion would be known perfectly through the premises and the premises through the conclusion. This opinion was rejected by Aristotle in the same text in no uncertain terms.

[2] The second opinion is that of others who, following Avicenna, took demonstration of the fact from its middle position[4] and so denied that there could be a demonstrative regress. Their basis was that in

demonstration of the fact one either begs the question[5] or proves nothing with necessity. This opinion was rejected by us in the first question.

[3] The third opinion is that of many moderns, whose leader was [Ugo Senensis],[6] who were opposed to the regress because a demonstration of the fact achieves everything offered by a demonstration of the reasoned fact. To the contrary: demonstration of the reasoned fact is most perfect and differs essentially from demonstration of the fact; therefore demonstration of the fact cannot offer the same result as demonstration of the reasoned fact.

[4] The fourth opinion is that of Franciscus Neritonensis,[7] whom many follow, who denied a second progression[8] in demonstration. For, seeing that there are two progressions in it, one from an effect and the other from a cause, he admitted the first but denied the second, because in the second progression, since we proceed from what is less known to us, we cannot infer a conclusion from necessary premises. To the contrary: as Aristotle teaches in the second book of the *Prior Analytics*,[9] every syllogism must proceed from things more known to us; therefore, either Neritonensis's[10] argument is invalid or the second progression must be admitted. Second: this opinion does away with demonstration of the reasoned fact, contrary to Aristotle and to common opinion.

[5] The fifth opinion is Aristotle's,[11] as above, denying that there can be a perfect circle in demonstration and yet admitting an imperfect one. This opinion we regard as true. For its understanding:

[6] Note, first: two things are required for demonstration. First, that the proving part and the part proved[12] must be connected with each other, for otherwise one cannot make a necessary inference from the one to the other. Second: that the proving part, as more known, should come first in the demonstration.

[7] Note, second: cause and effect can be taken in three ways: one way, under the formal relationship of cause and effect;[13] in a second way, as they are disparate things;[14] and in a third way, as the cause is necessarily connected with the effect.[15]

[8] These two notations presupposed, I say, first: if cause and effect are taken in the first way, there cannot be a demonstrative regress. Proof of the conclusion: demonstration must proceed from what is more known; but correlatives are so interdependent that one cannot be more known than the other;[16] therefore [there cannot be a demonstrative regress if cause and effect are taken as formally related to each other].

[9] I say, second: if cause and effect are taken in the second way, there

is no circularity.[17] The reason for this is that in a demonstration one thing must be inferred necessarily from the other; but things that are not necessarily connected with each other are so related that one cannot be inferred from the other; therefore [no circularity is involved if cause and effect are taken as disparate things].

[10] I say, third: since the demonstrative regress is a progression of reasoning in demonstration,[18] which is made from effect to cause and vice versa for the more perfect development of the sciences,[19] it can occur in cause and effect considered in the {third}[20] way, provided that it is in a different genus of causality, or even in the same genus when not considered under the same formality and not to the same term.[21] The reason for this conclusion: it can happen that one would know an effect and yet not know the cause, and consequently one might prove the existence of the cause from the existence of the effect. Again, it can happen that a person would discover a cause and be unaware of the proper way it produces [its] effect;[22] the proper way might then be manifested through the demonstrative regress. [A regress of this kind can be made when][23] there is a necessary connection, such as the reason why the effect accompanies the cause, and one of these can be taken as more known in order to prove the other; therefore [there can be a demonstrative regress when there is a necessary connection between cause and effect and under the proper circumstances].

[11] You object: it can happen that one and the same thing be the cause of another thing in the same genus of causality, and vice vera; therefore there can be perfect circularity.[24] Proof of the antecedent: vapors are the material cause of rain, and rain is the material cause of vapors; therefore [vapors can be the material cause of rain, and vice versa, so there will be perfect circularity]. I reply, first, by denying the inference: in a demonstration of this kind there is no progression to the same numerical term;[25] for although we may prove vapors through rain, when we then prove rain through vapors we do not show numerically the same rain to be involved; therefore [in such a case there is no perfect circularity]. I reply, second: a proof of this kind is not made in the same genus of causality, because when we prove rain through vapors, we demonstrate such rain in the sense that the vapors are able to be condensed; on the other hand, when we prove vapors through rain, we demonstrate such vapors in the sense that the rain is able to be rarified. As a consequence the middle term is not formally the same, and no circularity is involved.

[12] You ask, first: can the existence of the effect be shown in the

second progression[26] wherein the cause gives the proper reason for the effect? I reply, with St. Thomas: its existence can be proved through demonstration of the reasoned fact, though this is not absolute and simply perfect existence.

[13] You ask, second: in what sciences do we think such circularity is useful? I reply: the demonstrative progression is useful for the perfecting of all the sciences,[27] but it is most frequently used in physics because for the most part physical causes are unknown to us. In mathematics it has almost no use, because in such disciplines causes are more known both with respect to nature and with respect to us.

[14] You ask, third: what are the conditions for the demonstrative regress[28]? I reply: these are [six in number. First:][29] that there be two progressions of demonstration in it,[30] one from effect to cause, the other from cause to effect. Second: that we begin with demonstration of the fact, as Aristotle teaches, for otherwise demonstration of the fact would be pointless, since one who knows the proper reason for the effect also knows its existence, as we have explained.[31] Third: that the effect be more known to us, as Aristotle teaches and as is obvious to reason, for otherwise we could not formulate a demonstration of the fact. Fourth: that having made the first progression we do not begin the second progression immediately, but wait until we come to have formal knowledge of the cause we first know only materially.[32] The reason: because we cannot formulate a demonstration of the reasoned fact if we do not have prior formal knowledge of the cause. You object: therefore it would follow that demonstration of the reasoned fact would be made to no purpose, because it is made only that we may have formal knowledge of the cause.[33] I deny the inference: because, granted that one who has formal knowledge of the cause[34] also has virtual knowledge of the reason why the property inheres in the subject, one does not understand it actually unless one effects a true demonstration. And from this it follows that the regress is circular in an improper sense, since in it one progresses from an effect to material knowledge of the cause, and then from formal knowledge of the cause[35] to the proper reason for the effect. Fifth condition: that it be made in convertible terms, because if the effect were broader in scope than the cause, this would impede the first {progression}.[36] Hence the following is not valid: there is light, therefore the sun. If, on the other hand, the cause were broader in scope than the effect, this would impede the second progression, as is obvious. For, although it is valid to argue: something breathes, therefore it has a soul, one cannot do

the reverse, because respiration requires organs that may be lacking in things that nonetheless have a soul. Finally: the reasoning must be in the first figure.

NOTES AND COMMENTARY

D1: *On the Nature and Importance of Demonstration*

D1.1: On the definition of demonstration. *For background, see Secs. 2.5, 3.1–3, and 4.4 of* Galileo's Logic of Discovery and Proof, *for applications, Secs. 5.5 and 6.8*

[1] The title of the treatise is not set off on a separate line, though Galileo begins his exposition with the words *Tractatio de demonstratione*, obviously intended to serve as a title but becoming instead the subject of his first sentence.

[2] *quite judiciously [...]*: Lat. *sapientissime*, followed by a blank space of a third of a line. The space may be traceable to Galileo's having had difficulty reading the exemplar from which he worked; alternatively, he may have read it correctly but left room for a more precise reference to be filled in later. Vallius's version of 1622 has *in principio primi libri* at the ellipsis, i.e., "at the beginning of the first book," which would be a plausible reading [VL2: 188]. In the notes of Vitelleschi, who taught the logic course at the Collegio in 1589 (one year after Vallius), "texts 5, 6, and 8 of the first book" are cited in a similar reference [Lat. Ed. 175]; perhaps this is what Galileo was seeking to put in the space left blank in his manuscript.

[3] *about which we will have much to say in the following treatise*: a clear reference to a Treatise on Science and thus an indication that, at the time he was writing this, Galileo planned to appropriate from Vallius's notes an additional treatise on that subject. No such treatise is now extant, and whether or not Galileo persevered in his intention remains problematical. In either event, however, he must have read through Vallius's treatise on science and decided that it too was worth preserving among his notes. Since throughout the manuscript on which we are commenting Galileo evidences good knowledge of Jesuit teachings on science, the missing treatise is probably the source on which he drew. Much of the material presented in Secs. 3.1 through 3.7 of *Galileo's Logic of Discovery and Proof* is summarized from Carbone's plagiarization of the treatise in his *Introductio in universam philosophiam*; for details, see the Introduction to the translation above.

[4] *First Disputation*: Here Galileo simply wrote *Disputatio prima* and then, on the same line, began the first question without indicating the title of the disputation as a whole. Actually the disputation, as revealed in the description given in the prologue, is made up of two questions, the first on the definition of demonstration and the second on its importance (*praestantia*) when compared with another instrument of knowing, namely, definition.

[5] *many species of demonstration*: for previous mentions of the various kinds of demonstration, see F2.3.2 n. 6, F2.3.4 n. 9, F3.5 n. 2 and F4.1.3 n. 6; a fuller discussion of the different types will be found in D3.1 and its accompanying notes. See also Secs. 4.4a and 4.9a of *Galileo's Logic of Discovery and Proof*.

⁶ *ostensive demonstration*: Lat. *demonstratio ostensiva*, a new type that has not been noted previously. It proves a conclusion in a positive way, showing why something is the case rather than arguing indirectly by eliminating other possibilities. It may be viewed as a quasi-genus that includes all the other types except demonstration to the impossible (or to the absurd); the latter was commonly regarded as imperfect (F2.3.2) and thus was not of major interest for determining the nature of demonstration.

⁷ *true and proper principles*: see F2.1.1 n. 5.

⁸ *proper and proximate principles*: see F2.1.3 n. 9.

⁹ *universal and particular, affirmative and negative*: this division is mentioned later in D3.1.2, where it is identified as a division of demonstration as it is a syllogism.

¹⁰ *a type of illative discourse*: Lat. *discursus quidam illativa*, that is, a kind of reasoning process by which one discerns how a conclusion follows from one or more propositions, thus an argumentation; see D3.1.2.

¹¹ *an instrument of scientific knowing*: Lat. *instrumentum sciendi*, a broader category than discourse, embracing all acts of the mind whereby one knows or comes to new knowledge that is true and certain. The various instruments are enumerated and discussed in detail in the following question (D1.2). Vallius apparently composed an entire treatise on instruments of knowing, which was appropriated by Carbone and thus is preserved in the *Additamenta* of 1597; see Sec. 4.4c of *Galileo's Logic of Discovery and Proof*.

¹² *the possible intellect*: Lat. *intellectus possibilis*, the part of the human intellect in which concepts are formed when actuated by an appropriate stimulus deriving from the agent or "acting" intellect, the *intellectus agens*. Galileo's use of intellection (*intellectio*) here thus refers to the formation of concepts, the beginning of rational knowledge. See D1.2.2 n. 4, D1.2.6 n. 12; for a fuller account of the theory of knowledge on which this is based, see Secs. 2.1–2.3 of *Galileo's Logic of Discovery and Proof*.

¹³ *actual science...habitual science*: Lat. *actualis...habitualis*, a distinction that figures prominently in Jesuit treatises on science. Actual science designates the act of knowing that results from the intellectual grasp of a single demonstrative syllogism; habitual science, on the other hand, designates the habit of mind that results from repeated acts of this kind or the body of knowledge built up from them. A total science such as physics or metaphysics would be an example of habitual science, as would the partial sciences associated with them. See Sec. 3.2 of *Galileo's Logic of Discovery and Proof*.

¹⁴ *syllogism producing science*: Lat. *syllogismus faciens scire*, where *scire* means to know simply or in an unqualified way, that is, to know scientifically.

¹⁵ *material species...formal species*: material species would be various subject matters that fall under a genus without regard to the formalities that differentiate them, whereas formal species would be those distinguished by their own proper differentiae.

¹⁶ *instrumentally...dispositively*: in the sense that, although the intellect is the proper cause that produces science, the demonstrative syllogism is either its instrument or a necessary condition for its doing so.

¹⁷ *perfect science*: Lat. *perfecta scientia*, as opposed to the imperfect science that results from a demonstration to the impossible; see F2.3.2 n. 6, D2.1.5, D3.1.4.

¹⁸ *superior*: Lat. *nobilior*, literally more noble, to be taken in the sense of better or more excellent. The opposite conclusion, namely, that knowledge of the premises is superior to knowledge of the conclusion, is defended in the following question, D1.2.

¹⁹ *second definition*: with regard to the brevity of the explanation given here for the second

definition, it should be noted that each part of the definition is examined with care and in considerable detail in the second disputation of this treatise, D2, as follows:

[20] "of true premises": see D2.1.

[21] "of first and immediate": see D2.2 through D2.5.

[22] "of more known": see D2.6.

[23] "from antecedents and causes of the conclusion": see D2.2. Additional elaborations relating to the self-evidence, necessity, and the properties of the propositions that compose a strict demonstration are provided in D2.7 through D2.12.

D1.2: Is demonstration the best of all instruments of knowing? *For background, see Secs. 2.1–2, 2.5, 3.4c, and 4.4 of* Galileo's Logic of Discovery and Proof, *for applications, Secs. 5.2, 6.2c, and 6.8c*

[1] *best*: Lat. *nobilissimum*, literally, most noble, also translated below as most superior, most important, etc.

[2] *instruments of scientific knowing*: see D1.1.2 n. 11.

[3] *natural...adventitious*: Lat. *naturalia...adventitia*.

[4] *intellectual light*: Lat. *lumen intellectuale*, the light deriving from the agent intellect, which, in a Thomistic theory of knowledge, illuminates the phantasm and yields the intelligible species that, when impressed on the possible intellect, produces the concept. See D1.1.2 n. 10 and D3.1.17 n. 38; also Sec. 2.1a of *Galileo's Logic of Discovery and Proof*.

[5] *mediately...immediately*: for this couple Vallius-Carbone substitute "remotely... proximately" [CA24r]; the sense is that division and method aid in acquiring knowledge but do not actually supply it; other instruments supply knowledge directly, either imperfectly, i.e., without certitude or evidence (though they prove useful for discovering principles, or when demonstrations are unattainable), or perfectly, i.e., with certitude and evidence.

[6] *proposition*: treated more fully in paragraph [4] below.

[7] *method*: Lat. *methodus*, a term not explained by Galileo in the notes that are extant. Vallius treats it in his commentary on the second book of the *Posterior Analytics*, in his disputation on definition. For him, method is a procedure for ordering a discipline or a science so that each consideration of its subject matter has its proper place; understood in this way it is an instrument of scientific knowing distinct from all others. Following Galen, Vallius notes, method has three species, that of composition, that of division, and that of resolution [VL2: 395]. See Secs. 1.3 and 2.6 of *Galileo's Logic of Discovery and Proof*.

[8] *resolution and composition*: the process whereby a conclusion, as an effect, is "resolved" or analyzed back to the cause that produces it by *a posteriori* reasoning (the method of resolution), and then is "composed" or synthesized with the cause to establish the same conclusion by *a priori* reasoning (the method of composition). This twofold process, known among Paduan Aristotelians as the demonstrative regress, is explained and defended at length by Galileo in D3.3. See Secs. 1.3, 2.7, and 4.9 of *Galileo's Logic of Discovery and Proof*.

[9] *{definition}*: Galileo wrote "division" here, obviously a slip of the pen, as is discernible from the context.

[10] *order*: Lat. *ordo*, a term likewise not explained by Galileo. Vallius defines it as habit of mind by which one can arrange the parts of a discipline so as to proceed from the more known to the less known and thus can facilitate its being learned by others [VL2: 395–396].

Very similar to method, order is mentioned here as the genus under which method is contained; thus one can also speak of an order of resolution and an order of composition [*ibid.*]. See Sec. 2.6 of *Galileo's Logic of Discovery and Proof.*

[11] *by the natural light*: Lat. *lumine naturali*, that is, by the light of the intellect, according to the theory of knowledge that lies behind MS 27. See Secs. 2.1–2.3 of *Galileo's Logic of Discovery and Proof.*

[12] *intrinsic or extrinsic causes*: intrinsic causes are internal to the entities they cause, and thus formal causes and material causes are classified as intrinsic; extrinsic causes, on the other hand, are external to the entities they cause, and so efficient causes and final causes are classified as extrinsic; the distinction is employed at D1.2.25, D1.2.34, D2.8.4, D2.10.6, D3.1.14, D3.1.18. See also D2.2.4 and Sec. 4.5 of *Galileo's Logic of Discovery and Proof.*

[13] *operations of the intellect*: Lat. *operationes intellectus*, three in number and thus spoken of in scholastic treatises as the three acts of the mind. The first operation is that of simple apprehension, the grasping of concepts, as when one discerns immediately the meanings of terms; the second is composing or dividing, as when one uses concepts to form a judgment, expressed as a positive or negative proposition; and the third is reasoning or discourse, as when one arranges propositions in the form of a syllogism to arrive at a previously unknown conclusion. See Sec. 2.2 of *Galileo's Logic of Discovery and Proof.*

[14] *intelligences*: that is, intellectual substances or angels. They attain knowledge directly or intuitively, without need of discourse; in this they are different from rational beings or humans, who reason their way to new conclusions. See F3.1.16 n. 41; also Secs. 2.7a and 3.1b of *Galileo's Logic of Discovery and Proof.*

[15] *constitutive of a thing*: in the sense of being part of its definition; man is defined as a rational or discoursing animal, and thus discourse is part of his being.

[16] *[a] the superior and [b] the more extensive object*: Lat. *nobilius obiectum et latius patens*, a cryptic expression on which light is shed by Vallius-Carbone. They divide the objection into two parts, the first arguing that the superior instrument is concerned with the superior object, the second that the superior instrument makes the user know more things and helps in more ways, which would be equivalent to its having a more extensive object [CA28vb– 29ra; Lat. Ed. 189]. See D1.2.33 n. 31.

[17] *quiddity*: essence or definition, the answer to the question "What is it?" (*quid sit*); see F3.1.4 n. 16.

[18] *the "what it is"*: Lat. *ipsum quod quid est*, another way of referring to the essence or definition, actually a more literal translation of the expression used in text 24 of the original Greek [76b6]; for similar translations, see F3.6 n. 2 and F3.6.3 n. 7.

[19] *as a middle term, not as an instrument of knowing*: Galileo does not explain this distinction, but it is explained by Vallius-Carbone [CA25vb]. Their text is translated and located in the context here under discussion in Sec. 2.5b of *Galileo's Logic of Discovery and Proof.*

[20] *many moderns*: Lat. *multi recentiores*, unnamed by Galileo and Vallius-Carbone, but possibly Angelus Thius and Eustratius, identified by Vallius in a similar context [VL2: 407].

[21] *in itself and in its nature, or with respect to us*: Lat. *secundum se et suam naturam, vel respectu nostri*; see D2.6.1.

[22] *this difficulty*: the antecedent of the "this" is not clear. Possibly Galileo had in mind the problem posed in the title to the question, but Vallius-Carbone make explicit that the precise difficulty they are addressing is that posed by the third opinion. They then introduce this and

the other notations corresponding to Galileo's paragraphs [17] and [18] with the remark that their solution to it will require a number of notations [Lat. Ed. 191].

²³ *without movement...involves movement*: Lat. *sine motu...coniuncta cum motu*. The movement referred to here is that of the intellect from premises to conclusion in the process of demonstration.

²⁴ *correlatives*: Lat. *relativa*; see D3.3.8 n. 16.

²⁵ *something simple*: Lat. *quid simplex*, simple in the sense that an essence or definition is grasped directly in the first act of the mind and does not require composing and dividing, as does the second act, or discourse from premises to conclusion, as does the third. Further light on the conclusion stated in this paragraph and on the arguments given in its support is given below in paragraph [37]. See Sec. 2.2 of *Galileo's Logic of Discovery and Proof*.

²⁶ *procreated to attain beatitude*: the argument here presupposes that one can define man using all four causes, as follows: man is a being created by God and procreated by parents (efficient cause), composed of an animal body (material cause) and a rational soul (formal cause), to attain beatitude in heaven (final cause). Since the final cause is the cause of causes, and causes stand to each other in a definite hierarchy, with final, efficient, formal, and material being each superior to that following it, it would seem that one can demonstrate man's nature from his final cause.

²⁷ *from extrinsic causes*: that is, from the efficient cause (procreated) and the final cause (to attain beatitude), as in the preceding note.

²⁸ *without {movement}*: for *sine motu* here Galileo wrote *sine mora*, without delay. The reading makes sense, for movement takes time and thus to grasp something without movement would be equivalent to grasping it without delay. His reading has been emended here, however, to agree with his previous usage in D1.2.18.

²⁹ *{barely}*: Lat. *vix*. Galileo apparently had difficulty reading the exemplar from which he worked, first writing *vix*, then crossing it out and writing *bix* (which is meaningless) on the same line after it. Vallius-Carbone preserve the correct spelling [CA30rb].

³⁰ *primary analogate...analogy of proportion and also of attribution*: a terminology deriving from Thomas Aquinas. Analogy of proportion may be referred to as four-term analogy that takes the form: A is to B as C is to D. For example, a sensation (A) is to a sense power (B) as a concept (C) is to the power of intellect (D); in this sequence, it should be noted, no one term is said to be superior to the other. Analogy of attribution, on the other hand, may be referred to as two-term or three-term analogy wherein one term, the primary analogate (A), enjoys a primacy over another (B) or others (B and C) because of the ways in which they are related. For example, the health of a human organism (A) is the primary analogate to which healthy food (B) is related as a cause and a healthy complexion (C) as a sign. In this type of analogy food and complexion are said to be healthy only by way of attribution to the organism that is healthy.

³¹ *To the third*: Galileo lost track of the arguments here, for his reply is obviously to the fourth objection, paragraph [11], and not to the third, paragraph [10]. He further confounds the issue by reversing the two parts of the fourth objection when proposing his answer, replying first to [b] and then to [a].

³² *To the fourth*: the reply here is to the third objection, paragraph [10], the fourth having been answered in paragraph [33].

³³ *To the fifth*: Galileo's response here is somewhat cryptic. It can be elucidated by comparison with that of Vallius-Carbone [CA31ra, Lat. Ed. 199], which translates as

follows: "We concede that [a definition], as it is put in a demonstration, is a part and on that account is inferior, but the problem here is not about definition as it is considered in this way. That it is not put [in demonstration] as an instrument of knowing, as is maintained in the objection, is obvious." Note that Galileo, in his paragraph beginning "You say:" immediately following his fifth objection (D1.2.13), had already explained, and thus made obvious, why it is that definition is not put in demonstration as an instrument of knowing; Vallius-Carbone followed the same procedure in their parallel account.

³⁴ *To the sixth*: there is no sixth objection in Galileo's text, which should have been given immediately after paragraph [12] above. As pointed out in the Lat. Ed. 200, what Galileo is doing here is actually continuing the response that should have been given to the fifth objection, which he possibly realized he had shortened too much. To see this, all one need do is insert a "For" before the expression "definition as it occurs in demonstration" in Galileo's reply and then add his entire response to the response given by Vallius-Carbone and cited in the previous comment. This addition clarifies completely the otherwise cryptic response to the fifth difficulty. Why Galileo inserted "To the sixth" here is not clear, since he had only formulated five objections in paragraphs [8] through [12]. From the fact that Vallius-Carbone do record a sixth objection it seems probable that there was a sixth reply in Galileo's exemplar, and he inadvertently conflated this with what should have been the reply to his fifth objection.

³⁵ *You say*: The point of this additional argument is not clear, since it is not directly related to the material preceding it, nor does it seem to be a part of the missing sixth objection as this can be discerned from the material preserved by Vallius-Carbone. Light is shed on it, however, by the parallel passage in Vallius-Carbone [CA31rb, Lat. Ed. 200–201], who propose it as a further clarification of Galileo's second conclusion and the arguments offered in its support in paragraph [22]. Their observation translates as follows: "But a scruple here remains against what we said in the response to a difficulty made against the second conclusion, where we taught that a definition depends only instrumentally on the syllogism whereby the definition is investigated." They continue on, in terms similar to Galileo's: "For an instrument, once it is acquired or when the thing for whose doing it was needed has been done, is no longer necessary, as is apparent by induction: for a pen is no longer needed when letters are written. But the syllogism wherein the definition was found is always necessary, at least virtually, if we are to be said to know the quiddity with certitude. Therefore the definition does not depend on the aforesaid syllogism only instrumentally but also essentially, just as the conclusion depends on the premises."

³⁶ *I reply*: Galileo's reply here parallels the slightly fuller response in Vallius-Carbone [CA31rb, Lat. Ed. 201], which translates as follows: "Here the scruple is removed as follows: the major proposition is true of material instruments, but not of rational and intellectual ones. The reason for the disparity is that, for anyone to be said to know something perfectly it is necessary for him to know that he knows, and this cannot occur unless knowledge of the instrument of knowing remain at least virtually. An example may be taken from one who knows a thing certainly through some particular demonstration: he would not be said to know perfectly if he were not to possess, at least virtually, the teaching on demonstration in general, which is an instrument of knowing." Otherwise the teaching is the same in both versions. Which treatment duplicates more faithfully the original exemplar, presumably Vallius's lecture notes of 1588, is difficult to say. Since Carbone was known for his pedagogical skills, and Vallius complained about his rearranging the materials

of the lecture notes, it is quite possible that Galileo's version reflects obscurities that were present in the original and were later removed by Carbone; if so, Galileo's version would follow more closely Vallius's notes.

D2: *On the Properties of Demonstration*

D2.1: Is demonstration composed of true premises? *For background, see Secs. 2.3, 4.4, and 4.7 of* Galileo's Logic of Discovery and Proof

[1] *Properties*: Lat. *proprietates*, understood in the sense of the necessary conditions or qualifications that must characterize the propositions that make up a demonstrative syllogism.

[2] *of true premises*: Lat. *ex veris*, literally, of truths, with obvious reference to the truths contained in the premises.

[3] *three difficulties*: the first is stated in this paragraph, the second in paragraph [6] and the third in paragraph [9] below.

[4] *truth is twofold*: Vallius further notes that this teaching derives from Ammonius, Boethius, and St. Thomas Aquinas, and furnishes citations for each [VL2: 222, Lat. Ed. 205].

[5] *real*: Lat. *realis*. Real truth is the truth associated with the transcendental property of being, usually expressed in the axiom *ens et verum convertuntur*; on this account it is also referred to as ontological, or entitative, or transcendental truth.

[6] *of reason*: Lat. *rationis*. Truth of reason is truth in the ordinary sense, as when applied to a statement or to knowing of any kind; on this account it may also be referred to as epistemological truth or psychological truth.

[7] *simple...complex*: some commentators refer to simple or non-complex truth as material truth and to complex truth as formal truth; in his *Logica* [VL2: 222] Vallius speaks of the first as passive truth, the second as active truth. See Sec. 2.2b of *Galileo's Logic of Discovery and Proof*.

[8] *first operation...second and third operation*: see D1.2.6 n. 13 and Sec. 2.2 of *Galileo's Logic of Discovery and Proof*.

[9] *accidentally and not essentially*: Lat. *per accidens et non per se*, that is, indirectly or incidentally and not directly and from themselves.

[10] *from principles...related essentially*: Lat. *ex principiis per se*.

[11] *can be inferred...cannot be known scientifically*: Lat. *posse inferri, non tamen sciri*, that is, not known in the probative way proper to a *scientia*.

[12] *proof*: Lat. *probatio*. Vallius further explains that the inference or the illation is made in virtue of the form of argument, whereas scientific knowing results from its matter; thus the premises must not only be arranged in proper form but they must also contain true matter [VL2: 222–223]. See Sec. 2.7 of *Galileo's Logic of Discovery and Proof*.

[13] *ostensive demonstration*: see D1.1.1 n. 5.

[14] *in a qualified sense*: Lat. *secundum quid*.

[15] *to the person*: Lat. *ad hominem*.

[16] *non-complex*: Lat. *incomplexa*, the same as simple truth, equated by Vallius in his *Logica* with transcendental or entitative truth [VL2: 223].

[17] *real being*: Lat. *ens reale*, that is, being that is predicamental and knowable, and so truly being, for only this can be truly and properly grasped in a scientific way. It is opposed to rational being, mentioned in paragraph [8] and in the following note.

[18] *of rational being*: Lat. *de ente rationis*. Here Galileo implies that logic, whose object is rational being, cannot be a science; in F3.1.11 he qualifies this by noting that it suffices for the objects of rational sciences to have an "objective existence" in the mind as opposed to a real existence outside the mind. See F3.1.11 n. 33. and Sec. 2.3 of *Galileo's Logic of Discovery and Proof*.

[19] *a rose [in winter]*: Lat. *rosa in hieme*. In abbreviating his material Galileo omitted the *in hieme*, which was undoubtedly in his source and is translated here in brackets; it is preserved in Vallius's *Logica*, VL2: 223.

[20] *[to be in the thing]*: in Galileo's manuscript there is a lacuna after the words "to be known" (*cognosci*), a possible indication that Galileo had difficulty reading his exemplar and so left a space to be filled in later. The words omitted were *rei inesse*, as can be seen in Lorinus's exposition of Cajetan's teaching [LL464; Lat. Ed. 207]; they are translated here in the bracketed expression.

[D2.2]: Must premises of demonstrations be first and prior? *For background, see Secs. 2.7 and 4.2 of* Galileo's Logic of Discovery and Proof, *for applications, Secs. 5.2, 6.2c, and 6.8c*

[1] *from premises that are first and prior*: Lat. *ex primis et prioribus*, literally "from firsts and priors," with obvious reference to premises or propositions.

[2] *though formally different from each other*: the formal difference is that "first" means that nothing has preceded the premises, whereas "prior" means only that something comes after them regardless of whether or not anything has gone before.

[3] *more known...with respect to nature...to us*: see F4.2 for background; also D2.3.8 and D2.6 for a related discussion.

[4] *most perfect science*: Lat. *perfectissima scientia*. Since God is the most perfect being, the science that has him as its object should be the most perfect science.

[5] *there are no causes in him*: in the sense that God is the First Uncaused Cause, causative of all other things but not caused in any way himself. Still, causes can be said to be in God, as explained in paragraphs [2] and [6] below.

[6] *Note, first: causes in being*: this notation is formulated to prepare the answer to the second difficulty formulated in paragraph [1] concerning the existence of causes in God. In his *Logica* Vallius explains that the distinction between causes in being (*in essendo*) and causes in knowing (*in cognoscendo*) originated with Themistius, but that causes in being were further developed by the Thomistic commentator, Thomas de Vio Cajetan, to include virtual causes as well as those that are real and formal [VL2: 240].

[7] *Others are imperfect*: imperfect causes, here opposed to those that are true and proper and in paragraph [7] to those that are proper and perfect, are also known as virtual and verisimilar causes; see paragraphs [3] and [7], also F2.3.3 n. 7.

[8] *from all species of cause*: Lorinus gives a full citation of authorities in support of this conclusion, including Alexander of Aphrodisias, Averroes, Albert the Great, St. Thomas, Walter Burley, Paul of Venice, and Agostino Nifo [LL490].

[9] *formal causes...are more intrinsic to the thing*: Galileo's explanation here is somewhat cryptic. A fuller elaboration is given by Vallius in a passage that translates as follows: "Of formal cause in this sense it seems that Aristotle was properly speaking in the text where, giving an example of the formal cause, he gives a definition of a triangle, of a right angle, etc., through the material cause; he also gives other examples using the efficient and final cause through which a thing is defined. Although he does give these examples, the true formal cause is not excluded [by them]. For we can demonstrate through all kinds of cause, and this will produce true scientific knowing, because we will know that on account of which the thing is, and that this is its cause, etc. And generally speaking, whenever we demonstrate through any cause that enters into the definition we can be said to demonstrate through the formal cause, insofar as the definition is reducible to the formal cause and thus includes all those particular causes that make up the definition. For in this way the same entity considered under different formalities can be said to be the efficient, or the final, or the material cause, and at the same time the formal cause" [VL2: 245].

[10] *from a supposition*: Lat. *ex suppositione*. Galileo here presupposes knowledge of this expression, which takes on considerable importance in view of his later uses of it when explaining his own scientific methodology. In his later version of Galileo's likely exemplar, Vallius traces the origin of the expression to St. Thomas Aquinas in his commentary on Aristotle's *Physics* (Bk. 2, lect. 15), where he explains the different types of necessity associated with demonstrations from formal and material causality [VL2: 244]. Vallius's exposition is translated in Sec. 4.2b of *Galileo's Logic of Discovery and Proof*, which may be consulted for fuller details.

[11] *first principles of mathematics*: for example, that a whole is greater than any of its parts, or that when equals are added to equals the results are equal.

[12] *higher kind of knowledge*: as opposed to science, the higher kind of knowledge of God referred to here would seem to be wisdom or theology, which investigates all things in terms of their first and ultimate causes.

[13] *virtual causes {are}*: for the "are" here Galileo unaccountable wrote "in," and thus the term is enclosed in braces.

[14] *but also in the order of being*: this is Cajetan's innovation, as noted in D2.2.2 n. 6.

[15] *a definition implicitly involves formal causes*: see Vallius's explanation of this statement in n. 9 above.

[16] *either actually or virtually*: although the ability to reason is not actually the cause of risibility, being able to reason nonetheless enables one to wonder and in this sense can be said to be the proximate, though virtual, cause of risibility. See the fuller analysis of the syllogism discussed in this paragraph in Sec. 2.7a of *Galileo's Logic of Discovery and Proof*.

[17] *only those grades of definition*: that is, "rational" and "animal" and these alone, as opposed to taking "rational" to include the *virtus*, or ability, to be also "discursive" and "wondering." See D2.11.5 and the comment in its note (n. 13).

D2.3: What does Aristotle mean by immediate propositions? *For background, see Secs. 4.2, 4.6, and 4.7 of* Galileo's Logic of Discovery and Proof, *for a possible application, Sec. 6.8b*

[1] *by immediate propositions*: Lat. *per propositionibus immediatis*.

[2] *[need]*: not in Galileo's text; supplied for sense.

³ *of these types*: that is, propositions that are first to a subject or first to a conclusion, examples of which are cited in paragraph [1], namely, "Man is risible" and "Man is a rational animal." The demonstration *a priori* of man's risibility from his rationality is given in D2.2.8, while the demonstration *a priori* of man's rationality from his ordination to beatitude is given below in paragraph [4].

⁴ *can be demonstrated in an improper sense*: that is, by an argument of the following type: "Contradictory propositions cannot both be true at the same time: but some men are not irrational; therefore all men are rational." This is not a proper demonstration in the sense that it does not give a cause or reason for man's being rational, but merely manifests it by negating the contradictory and applying the principle of the excluded middle.

⁵ *from certain of his properties*: that is, from his risibility, his being teachable, his being grammatical, his being mathematical, his capacity for responsible moral behavior, etc.

⁶ *procreated for beatitude*: see D1.2.25 n. 26 and n. 3 above.

⁷ *"from the nature of the thing"*: a literal translation of the Latin, *ex natura rei*. This distinction was much used by Scotists and served to differentiate them from Thomists such as Cajetan; it was also adopted by the Jesuit theologian, Francesco Suarez, which perhaps explains its use by Vallius in reply to Cajetan's argument. Thomists distinguished a real distinction from a rational distinction or distinction of reason, and then further differentiated each of these into two subtypes: the real distinction they divided into the absolute distinction (that of one thing from another, *res a re*) and the modal distinction (that of a thing from its mode, *res a modo eius*); the rational distinction they divided into the greater, the distinction of reason with a foundation in reality (that of reason reasoned about, *rationis ratiocinatae*), and the lesser, the distinction of reason without a foundation in reality (that of reason reasoning, *rationis ratiocinantis*). Suarez named the distinction favored by Scotus, the formal distinction from the nature of the thing (*distinctio formalis ex natura rei*), the "intermediate distinction," that is, a distinction midway between the Thomists' modal distinction and their greater distinction of reason, the distinction of reason reasoned about. The point of the distinction *ex natura rei*, unimportant for understanding Galileo, gives some appreciation for why Scotus came to be known as "the Subtle Doctor" within scholasticism. A fuller discussion of the problem, with appropriate references, is to be found in Vallius [VL2: 229–232].

⁸ *this begs the question*: Galileo writes this in the infinitive form (*peti principium*). His interest in this expression has already been noted in F3.5.3 n. 7.

⁹ *one category [of being] is denied of another*: for example, in the propositions "Quality is not substance," "Quantity is not relation," etc. Since the categories are the supreme genera of being, no middle term is available to come between them, and thus the negation of their identity must be immediate.

¹⁰ *one differentia is denied of another*: for example, "Rational is not irrational," where the two differentiae would be applied within the genus "animal." or "Discrete is not continuous," where the differentiae would be applied within the genus "quantity," for reasons similar to that given in the previous note.

¹¹ *divides immediate propositions otherwise*: that is, on the basis of immediacy of assent to them rather than on the immediacy of their terms – stressing "with respect to us" rather than "with respect to nature," as in D2.3.8.

¹² *Axioms...Positions...suppositions*: Lat. *dignitates... positiones... suppositiones*. Because of the importance of these expressions for an understanding of Galileo's logical

methodology, a translation of Vallius's parallel exposition of them [VL2: 218] has been given in Sec. 4.2a of *Galileo's Logic of Discovery and Proof,* which should be consulted for fuller details.

[13] {*terms*} *or definitions*: Here Galileo wrote "principles or definitions," which is meaningless in this context; the correct expression, following Vallius as cited in the previous note, has "terms" instead of "principles," and thus "terms" is indicated here in braces.

[14] *self-evident*: Lat. *per se nota,* known on its own terms.

[15] *from the nature of the thing*: a repetition of the expression in paragraph [5], but this time apparently without the Scotistic connotation explained in n. 7 above.

D2.4: Must every demonstration be made from immediate premises?
For background, see Secs. 2.7, 3.3, 4.2, and 4.7 of Galileo's Logic of Discovery and Proof, *for applications, Secs. 5.1 and 6.6b*

[1] *from immediate premises*: Lat. *ex immediatis,* with obvious reference to premises or propositions.

[2] *most powerful*: Lat. *potissima,* similarly defined at F4.1.3. See also D2.12 and D3.1.

[3] *a proper attribute*: Lat. *propria passio,* as in F4.1.2 n. 5.

[4] *no less characteristic*: Lat. *non minus intrinsecum,* that is, no less pertaining to the nature or definition of demonstration.

[5] *should suffice*: Lat. *sufficiet;* an objection apparently directed against the plural form, *ex immediatis,* and thus implying that only one of the premises (*unica tantum*) need be immediate. In addition to the two objections raised here, [a] and [b], Vallius offers a third and more serious difficulty based on the teachings of Philoponus, St. Thomas, Apollinaris, Nifo, and others, who hold that true demonstrations can be made from mediate premises, against Averroes, who maintained that they must always be made from immediates. Vallius notes that the use of mediate premises is especially characteristic of the mathematical sciences, "for they prove a first conclusion in a first demonstration and assume it as a premise in a second demonstration, and the second conclusion as a premise in the third, and so on, and yet in mathematics these are true demonstrations" [VL2: 235–236].

[6] *in some way*: Lat. *aliquo modo,* an important qualification, very much needed if one is to reply to the additional objection noted in the previous comment. The qualification is further elaborated in paragraph [4] below.

[7] *subalternated science...proved in its* {*subalternating*} *science*: Lat. *scientia subalternata...probata in subalternante.* Here Galileo made an error, repeating *subalternata* at the second occurrence rather than writing the correct form, *subalternante.* Subalternation, as applied to sciences, is here understood to mean the subordination of one science to another by reason of its dependence on the other for its principles; the subalternating science, in such a case, is the one that supplies the principles on which demonstrations are based, whereas the subalternated science is the one that assumes these principles without proof and uses them in its demonstrations. Mathematical physics would be an example of a subalternated science, since it uses principles provided by mathematics (its subalternating science) to demonstrate conclusions in the realm of physics. Fuller details on the subalternation of the sciences are given in Secs. 3.3c and 3.4 of *Galileo's Logic of Discovery and Proof.*

⁸ *from a supposition and in a qualified way*: Lat. *ex suppositione et secundum quid*, an application of the teaching on supposition explained in D2.3.7C. Lorinus's fuller reply to the difficulty [LL477] translates as follows: "If anyone demonstrates through mediate principles without previously having their demonstration he will have science only from a supposition because he supposes these principles to be proved elsewhere. This is gathered from text 5 and from the last chapter of the second book of the *Posterior Analytics*, where it is said that one cannot know scientifically if one lacks previous knowledge of principles. And the reason is clear enough, because a conclusion is known only because of the principles, and therefore the latter must be known; and if they are known only from a supposition, then the conclusion is known only in this way also. From this it follows that demonstrations found in a subalternated science without those in the subalternating science are unable to generate perfect science." Again see Sec. 3.3c of *Galileo's Logic of Discovery and Proof*.

⁹ *from indemonstrables*: Lat. *ex indemonstrabilibus*, that is, from premises that are self-evident and thus need not, and cannot, be demonstrated.

¹⁰ *most perfect*: Lat. *perfectissima*, another way of speaking of a most powerful demonstration based on the fact that it generates a most perfect science. Vallius amplifies this reasoning as follows: "...a most powerful and most perfect demonstration should be such that it itself can infer a conclusion without any other and can fulfill in a perfect way all the conditions set down by Aristotle when treating of demonstration" [VL2: 236].

¹¹ *of virtually indemonstrable premises*: Lat. *ex indemonstrabilibus virtute*, the qualification implied in the first conclusion stated in paragraph [2] at n. 6 above.

¹² *immediate either actually or virtually*: Lat. *ex immediatis actu aut virtute*. Note here that the sense of immediate has been relaxed to include mediate premises, provided that the latter can be resolved to immediates. For similar uses of the terms *virtute* and *virtualiter*, see F2.3.3 n. 7 and D2.2.7.

¹³ *demonstration in general, taken analogically*: Lat. *demonstratio in commune sumpta analogice*, another Thomistic influence as noted in D1.2.29 n. 30. Analogy of attribution is probably intended here, with demonstration from premises that are actually immediate being the primary analogate, since only with reference to it can arguments from premises that are virtually immediate be said to be demonstrations.

¹⁴ *that serves as a connector*: Lat. *ratione cuius nectatur cum illis*, where the *cum illis* refers to the middle term or the subject. In the two examples that follow the first makes an addition next to the subject (*animal discursivum*, "discursive animal," to *homo*, "man") whereas the second makes an addition next to the middle term (*quoad animal rationale*, "as a rational animal," to *admirativum*, "capable of wonder"). See Sec. 2.7a of *Galileo's Logic of Discovery and Proof*.

¹⁵ *Every being capable of wonder is risible*: here Galileo first wrote "Every risible being is capable of wonder," and then interchanged the subject and the predicate in his final redaction.

¹⁶ *through the {subject's} definition*: Lat. *per definitionem illius*, literally "through the former's definition," if one were to follow the convention of translating *hic* and *ille*, and their inflected forms, as "latter" and "former" respectively. We here emend Galileo's *illius* to *huius*, as required for sense.

D2.5: Do all self-evident principles enter into every demonstration?
For background, see Secs. 2.7, 3.4, 4.2, and 4.7 of Galileo's Logic of
Discovery and Proof, *for applications, Secs. 5.3, 5.5, 6.1, and 6.2a*

[1] *[Fifth Question]*: neither an ordinal number nor the customary abbreviation for "question" are given in the manuscript, which begins immediately with the centered title: *An omnia principia immediata per se nota ingrediantur quamcumque demonstrationem?*

[2] *axioms*: Lat. *dignitates.*

[3] *foreknown to be true*: Lat. *praecognoscendum quia vera sint*; see F2.1, F2.2, and F2.3.

[4] *virtually*: see F2.3.3 n. 7.

[5] *in text 25*: that is, in Bk. 1, ch. 10, 76b13–77a9.

[6] *"Every whole is greater than its part"*: that is, a principle that is true from the meaning of its terms, and that is common to, or has application in, different sciences, such as arithmetic, geometry, and physics.

[7] *proper, immediate, etc.*: that is, principles that are first and immediate in a particular science and are not common to several.

[8] *texts 20 and 25*: text 20 is the whole of ch. 7, Bk. 1, 75a37–b21; text 25 is as given in n. 5 above.

[9] *common to all the sciences*: that is, principles that are regulative of all thinking and thus are not restricted to a particular subject matter.

[10] *restricted to some determinate subject matter*: although regulative of all thinking, principles such as that stated in paragraph [4] can be so formulated that they apply only, say, to numbers, as in "Every sum of positive whole numbers is greater than the positive whole numbers being summed."

[11] *mathematics*: where it might read "A number cannot be even and odd (i.e., not even) at the same time." Lorinus makes the comment that the foregoing distinctions are made mainly because of Aristotle's terminology in text 24 (Bk. 1, ch. 10, 76a32–b13), where all of his examples are taken from mathematics [LL479–480].

[12] *an imperfect demonstration*: for example, a demonstration that establishes a truth by showing the impossibility of its opposite; see F2.3.2 n. 6.

[13] *texts 24 and 26*: for text 24 see n. 11 above; text 26 is at Bk. 1, ch. 11, 77a10–26.

[14] *in mathematical demonstrations*: for example, one may prove that a particular line is longer than another by supposing that the first line is either longer than the second or it is not, and then by showing that one or the other of the alternatives leads to an impossibility and on this account must be false. In virtue of this reasoning, one is forced to conclude that the remaining alternative is true. The proof then requires that the principle stating that one of two contradictories must be true, as formulated in paragraph [5], enter actually into the demonstration. See D2.3.4 n. 4; also Sec. 3.4 of *Galileo's Logic of Discovery and Proof.*

[15] *such axioms*: that is, those that are immediately self evident, the subject of the first conclusion, paragraph [6] above.

[16] *any demonstration*: in view of paragraph [6], the sense would seem to require "any proper or perfect or ostensive demonstration," so as to exclude those that are improper or imperfect.

[17] *in themselves...in relation to our knowledge*: Lat. *secundum se... respectu cognitionis nostrae.* The distinction parallels that between "order of being" and "order of knowing"; see F2.3.4 n. 8. Vallius makes a similar distinction, though he expresses it in the form of an

objection: "You say: almost all mathematical conclusions depend on such universal principles. I reply: this is incidentally and with respect to us, and therefore we do not deny that knowledge of these principles is sometimes required; we deny only that they are always required" [VL2: 238–239].

[18] *an {imperfect} demonstration*: Galileo apparently made a copying error here, writing "perfect" where the sense obviously requires "imperfect."

[19] *a person who denies the proper principles of a particular science*: see the similar observation at F2.3.4 n. 9.

[20] *Here one may inquire*: as noted in the Latin Edition, 233, Galileo effectively begins a new question here, one with counterparts in both Lorinus and Vallius. Lorinus poses the question as follows: "Is it necessary to know all causes and to make resolution to the first in any genus?" [LL490]. Vallius, on the other hand, addresses the problem without putting it in the form of a question: "Chapter 8. It is shown that each demonstration is not to be resolved to first principles that are most common" [VL2: 239]. It is possible that the addition of this question was responsible for Galileo's losing track of his questions in this disputation, for the next three questions remain unnumbered in his manuscript and the title of one question [D2.8] is omitted entirely. See the prologue to Sec. 4.7 of *Galileo's Logic of Discovery and Proof*; also D2.8 n. 1 below.

[21] *resolved to first principles*: Lat. *resolvenda usque ad prima principia*. Galileo does not make clear what kind of first principles are intended here, those that are first in any genus, as Lorinus formulates the question, or those that are most common, as stated by Vallius. In his conclusions, however, he takes account of both. It is noteworthy that this is one of the few references in these questions to resolution, a term in Renaissance Aristotelianism usually coupled with composition, as in D1.2.3 n. 8.

[22] *in an unqualified way...in a determinate genus*: Lat. *vel simpliciter et absolute, vel secundum quid et in determinato genere*. Galileo was probably abbreviating heavily at this point, for he simply makes the distinction without giving any indication of how it is to be applied in what follows. Lorinus and Vallius use the distinction in their replies to the questions cited in n. 20 above. But see nn. 23–26 below.

[23] *I say, first*: apparently this conclusion relates to the first branch of the distinction made in paragraph [11], i.e., to scientific knowledge that is unqualified and absolute.

[24] *I say, second...its proper causes*: Galileo had a *lapsus* here, for in composing his notes he omitted the second conclusion as stated and later had to insert it in the margin of his manuscript. It obviously relates to the second branch of the distinction made in paragraph [11], i.e., to scientific knowledge in a qualified way and in a determinate genus.

[25] *Proof of the conclusion*: this sentence answers the question posed by Lorinus (see n. 20 above), for "proper causes" would be first in any particular genus.

[26] *But with respect to us*: similarly, this addresses the problem as formulated by Vallius (n. 20 above), for "first principles that are self evident" would be most common and most knowable for us.

D2.6: Is demonstration made from premises that are more known? *For background, see Secs. 2.6–8 and 4.7 of* Galileo's Logic of Discovery and Proof, *for applications, Sec. 6.2b–c*

[1] *[Sixth Question]*: This question, like the previous one, is unnumbered by Galileo; the numeration supplied here establishes continuity to D2.10, where Galileo resumes his numbering of the questions.

[2] *from premises that are more known*: Lat. *ex notioribus*, referring to the premises, as is clear from the clause that follows.

[3] *knowledge of the premises*: Lat. *cognitio praemissarum*.

[4] *better and more perfect than*: Lat. *maior et perfectior*, literally "greater and more perfect than." The sense of "greater than" (*maior*) is "more evident than" or "superior to," as becomes clear in Galileo's reply.

[5] *with respect to nature...with respect to us*: Lat. *secundum naturam ...secundum nos*, a distinction already invoked in D1.2.15.

[6] *confused knowledge...distinct knowledge*: Lat. *cognitio confusa... cognitio distincta*. Confused knowledge is knowledge that contains only a few distinguishing notes and thus is applicable to many things of different kinds; thus "animal" is a confused concept because it applies indifferently, i.e., confusedly, to many species of animal, whereas "squirrel" is distinct because it applies clearly and unambiguously to the squirrel species.

[7] *universals...in the order of causality...in the order of predication*: Lat. *universalia...in causando...in praedicando*. A universal cause would be one that is capable of producing a wide range of effects, as does the sun, for example, with respect to earth; other examples would be the first mover and protomatter, mentioned by Galileo in D3.2.3. A universal predicate would be one that can be predicated of a wide range of subjects, as "being," which applies to everything that exists, has existed, or can exist in any way whatever.

[8] *These distinctions understood*: that is, the distinctions just made in paragraph [2]. Galileo is not clear, however, on the way in which the two sets of distinctions are here being applied. Fortunately a parallel passage is available in Lorinus that casts light on Galileo's intended meaning: "These distinctions understood, I reply first that a universal in the order of causality, especially if it is a cause that is very common, would usually be more known because it is farther from the senses... Second, I reply that singulars properly speaking and individuals absolutely speaking are more known to us than universals... From these conclusions it will be easy to reconcile these two texts: for in the first book of the *Physics* Aristotle is speaking of universals that are more or less extensive in the order of predication, and which at the same time are universals in the order of causality and so are grasped in confused knowledge... In the first book of the *Posterior Analytics*, on the other hand, he speaks of the same universals grasped with distinct knowledge, since from their distinct grasp one comes to distinct knowledge of the conclusion" [LL482]. See Lat. Ed. 240.

[9] *Posterior Analytics, text 5*: i.e, in part of the text, at 71b30–72a6.

[10] *[in the order of predication]*: a qualification omitted by Galileo and here supplied for sense, following Lorinus's reading as in n. 8 above.

[11] *in the first book of the Physics*: in ch. 1, at 184a17–b14.

[12] *{and}*: in place of "and" (*et*) here Galileo wrote "this" (*hoc*), obviously a *lapsus calami*.

[13] *{indeed}*: in place of "indeed" (*quidem*) here Galileo wrote a word meaningless in this context (*quid*), apparently forgetting to add the *-em* that is required for sense.

¹⁴ *[to us]*: another omission that must be supplied for sense.

¹⁵ *demonstrations of this kind are not most powerful*: usually mathematical demonstrations employ such common principles that they are incapable of proving properties that are commensurately universal with their subjects and thus do not fulfill the strict requirements for most powerful demonstration; see also D2.12.8 n. 17.

¹⁶ *Concerning the second query*: Lat. *Circa secundum*, the reference being to the second half of the question in the title. The matter treated here has considerable affinity with that discussed in D1.2, which may be consulted for background.

¹⁷ *a natural cause that is not impeded*: a principle similar to that already used in F4.1.9, namely, "every natural cause that is sufficient to produce its effect does so necessarily as soon as the requisite conditions are provided." The "requisite conditions" referred to here would seem to include the removal of impediments to the causal action; for the importance of this teaching in Galileo's methodology, see F3.1.9 n. 27.

¹⁸ *if each [degree of] knowledge were gradually decreased*: Lat. *si paulatim minuatur utraque cognitio*: the "calculatory" expression *gradus cognitionis* is not used here, but that it is intended is clear from the reply in paragraph [16] below, where it is made explicit.

¹⁹ *is superior*: the Lat. for "superior" here is *nobilior*, literally, more noble.

²⁰ *[fourth]*: omitted by Galileo and thus enclosed in brackets.

²¹ *in subalternated sciences...in our faith*: for example, in mathematical physics, as subalternated to mathematics, the physicist may be certain of the principles he takes from the mathematician though they may not be evident to him; not being evident to him, he may be said to take them on faith. For a similar reason St. Thomas Aquinas regarded sacred theology as a subalternated science since it employs revealed principles accepted only on the basis of divine revelation and thus assented to by faith. See Secs. 3.3c and 3.4b of *Galileo's Logic of Discovery and Proof*.

²² *of first principles*: Lat. *primorum principiorum*; see D2.2.

²³ *[is called]*: Lat. *vocatur*, omitted by Galileo and here supplied in brackets.

²⁴ *of immediate principles*: Lat. *principiorum immediatorum*; see D2.3.

²⁵ *"that on account...more so itself"*: Lat. *propter quod unumquodque tale et illud magis*, a principle already stated in F2.1.3 and exemplified in its accompanying n. 10.

²⁶ *because of the slight regress*: Lat. *propter parvum recessum*. The point is that risibility is so closely related to rationality that one grasps its cause almost as soon as one understands what it is. In most cases where the demonstrative regress is employed, say, in the natural sciences, causes are hidden and a more extensive search or regress is required to discover them, as explained in D3.3. Even in the case of risibility, however, there are technical problems relating to its immediate cause, as touched on in D2.2.11 n. 16, D2.3.3 n. 3, and D2.4.6 n. 14. See Sec. 2.7a of *Galileo's Logic of Discovery and Proof*.

²⁷ *conclusion...principles*: the teaching here is obviously related to the theses advanced in D1.2, which may be consulted for background.

²⁸ *efficient cause*: Lat. *causa efficiens*. This would seem to be a further precision of the teaching in D1.1.2, where the intellect itself is identified as the efficient cause of demonstrative knowledge.

²⁹ *the major premise*: that is, "a natural cause that is not impeded produces an effect equal to it in perfection"; see n. 17 above.

³⁰ *a cause that is univocal*: Lat. *causa univoca*, opposed to a *causa aequivoca* or equivocal cause. In human generation man, in the sense of human parents, would be said to be the

univocal cause of man, in the sense of human offspring, whereas the sun, as also exercising causality in man's generation, would be regarded as his equivocal cause. Galileo uses this terminology in his *Tractatus de elementis* [GG1: 128, 165], where he also identifies the heavens as "a universal and equivocal cause" of sublunary events.

D2.7: Must demonstration be made from propositions that are immediate and said of every instance? *For background, see Secs. 2.7, 3.4, 4.2, and 4.8 of* Galileo's Logic of Discovery and Proof, *for a possible application, Sec. 6.6b*

¹ *Question [Seven]*: Galileo wrote the word "Question" before the title, but left a space for the number, apparently intending to fill it in later. The "Seven" is added here to provide continuity to D2.10, where Galileo resumes his own numbering of the questions.

² *from propositions that are necessary and said of every instance*: Lat. *ex propositionibus necessariis et de omni*. The *de omni* here translates literally as "of all" or "said of all," but among Aristotelian scholars it is usually translated as "said of every instance" to avoid difficulties that arise from the other types of universal predication considered in this question.

³ *unqualified*: Lat. *simpliciter*, with the connotation of being divine and uncreated. In a parallel passage Lorinus defines it as "what cannot be impeded in any way and consists in an actual existence that applies to God alone" [LL522].

⁴ *natural*: Lat. *naturalis*. In the same passage cited in n. 3 Lorinus refers to this necessity, which he says can be impeded by God, as created and natural (*creata et naturalis*).

⁵ *according to the ordinary law of God*: Lat. *secundum ordinariam Dei legem*, also referred to as "according to God's ordinary or ordained power," to differentiate it from the other branch of the division, *secundum absolutam illius potentiam*, "according to his absolute power." See Sec. 4.8a of *Galileo's Logic of Discovery and Proof*.

⁶ *[for the sun] to rise and set*: Lat. *solem oriri et occidere*. In writing this Galileo inadvertently left out the *solem*, here supplied for sense.

⁷ *which are necessary considered in their totality but not in their various parts*: in the sense that the element water, as a whole, necessarily is in its natural place between that of the elements earth and air, also considered as wholes, but a part of water, say a raindrop, need not necessarily be in that place but can be in air before falling back to its natural place.

⁸ *having an intrinsic order to each other...an extrinsic order to each other*: Lat. *habentia intrinsecum ordinem inter se...extrinsecum ordinem inter se*. An "intrinsic order" may alternatively be termed an "essential order" or a "necessary connection," whereas an "extrinsic order" may be termed an "accidental order" or a "contingent connection," as becomes clear in the subsequent development of the question.

⁹ *non-complex and complex*: Lat. *incomplexa et complexa*. For example, God's existence is necessary with a non-complex necessity, whereas the existence of a rose or an eclipse is necessary with a complex necessity, that is, on the supposition of there being places and times at which conditions are propitious to necessitate their existence, which requires a proposition being formed by the intellect for its very assertion.

¹⁰ *posterioristic statement*: Lat. *dictum posterioristicum*. In paragraph [3] below Galileo equates this with the *dictum per se* or the *propositio per se* [see n. 14]. The expression

propositio per se is translated in what follows as "essential proposition," thus equating *per se* with essential, as opposed to *per accidens* or accidental. Posterioristic statements therefore involve subjects and predicates that have an intrinsic order to each other, as explained in n. 8 above.

[11] *according to form...according to matter*: Lat. *secundum formam... secundum materiam.* On the basis of this distinction the scholastics differentiated formal logic from material logic, the first being based on the *Prior Analytics* and the second on the *Posterior Analytics.* See Sec. 2.7 of *Galileo's Logic of Discovery and Proof.*

[12] *"God is unchangeable"*: there is no plurality of subjects here, since the subject is singular; there are no inferiors that can be subsumed under God and to which the predication can be applied universally.

[13] *the commensurate or universal statement*: Lat. *secundum quod ipsum seu universale.* The expression *secundum quod ipsum* is the more or less literal Latin translation of the Greek expression *kath auto* at Bk. 1, ch. 4, of the *Posterior Analytics* (73b25), which is difficult to render into English; "precisely as such" captures some of the sense, but because of its connotations of universality it is usually translated as "commensurate" or "commensurately" universal.

[14] *every proposition that is essential*: Lat. *omnis propositio per se.* Note that here Galileo equates the posterioristic statement with the essential proposition, as remarked in n. 10 above.

[15] *every proper and perfect demonstration of the reasoned fact*: note that Galileo is maintaining here that a demonstration of the reasoned fact need not be made from propositions that are commensurately universal for it to be considered proper and perfect, but that it suffices for it to be made from propositions that are said of every instance and are necessary in the sense of being essential or posterioristic.

[16] *first and adequate subject*: that is, its proper subject, one that is equal in extension to the properties predicated of it; see also F3.2 n. 1.

[17] *with a [contingent] effect*: Galileo did not add the qualifier "contingent" in formulating the query, but the sense of the question requires it, since the effect's being contingent might call into doubt the necessity of the demonstration.

[18] *the ability to be eclipsed*: Lat. *eclipsibilitas*, eclipsability.

[19] *mathematicians [demonstrate] eclipses*: in writing this Galileo left out the word for "demonstrate" (*demonstrant*), which is required for sense. Note the use of "mathematician" to denote "astronomer" in our modern sense. See Sec. 3.4b of *Galileo's Logic of Discovery and Proof.*

[20] *from the middle term they use*: the middle term, in this case, would be the interposition of the earth between the sun and the moon, and since the earth is "true and real" this would be a good indication that the lunar eclipse it causes is itself real; see D2.8.4 n. 14. In his *Logica* Vallius formulates a similar query but answers it differently: "If someone were to object that an eclipse of the sun and the moon is demonstrated and that this is not always taking place, I reply, with text 65, that the eclipse is demonstrated insofar as it does always exist, namely, in relation to its proper causes" [VL2: 283]. (Text 65, in Vallius's enumeration, corresponds to text 22 in Galileo's, specifically, the passage at 75b33–37 in Bk. 1, ch. 8, of Aristotle's *Posterior Analytics.*)

[21] *said of every instance is threefold*: Galileo's explanation of this statement is somewhat cryptic and gives no examples; additional light can be shed on it from the following parallel

text in Vallius. "In the 'said of every instance' of which we are speaking we can assign three degrees. The first is in a certain way principal and should be said to be such absolutely and in an unqualified way: this 'said of every instance' is when the predicate is in each and every subject always, at all times and for all subjects, as when we say 'Man is an animal' or 'Man is rational' or anything similar. The second grade, less perfect, is when the predicate is always present at whatever time it is supposed to be there; of this kind are practically all astronomical propositions, as that the moon is being eclipsed, that Saturn is in conjunction with Venus, and others of this kind. The third is when the predicate can be truly present, and, if impediments that rarely occur are taken away, is always present; of this kind are practically all meteorological propositions, as that it will rain or snow at such and such a time, for although this takes place almost always it can nonetheless at some time be impeded" [VL2: 255]. Vallius's examples for these three grades illustrate the three kinds of "said of every instance" mentioned by Galileo.

[22] *Demonstrations can be made from all three of these*: Galileo does not explain here how these demonstrations can be formulated, but in the second and third types the demonstration would have to be made *ex suppositione*, namely, supposing the proper times and places, or supposing that impediments that might prevent the occurrence from taking place have been removed. See F3.1.7 n. 22, D2.2.5 n. 10, and D2.4.2 n. 8; also Sec. 4.2 of *Galileo's Logic of Discovery and Proof*.

D2.8: How many modes of speaking essentially are there? *For background, see Secs. 4.5 and 4.8 of* Galileo's Logic of Discovery and Proof, *for applications, Secs. 5.1, 5.2, 6.2b, 6.5b, and 6.8c*

[1] *[Eighth Question...contained under them?]*: enclosed in brackets to indicate that these words have been supplied by the editor. This entire title and its numbering as a question are missing in Galileo's manuscript. Apparently he started a new question here without realizing it, possibly thinking that he was continuing to reply to the "and how?" query in the second part of the previous question, D2.7. Such a *lapsus mentis* is difficult to explain. It may be that the title of the question was missing in the exemplar from which he worked; instances of this are found in some lecture notes at the Collegio Romano, where the copyist has left a space for a title to be lettered in later in another hand. If this were the case, then Galileo might have closed up the interval to save space in his own copy, not being aware when so doing that he was merging two questions into one.

The evidence that a new question is being begun and that a new title must be inserted here is twofold. First are the expressions "Concerning the first query" and "Concerning the second query," with which the first two paragraphs begin; these respond to two questions that appear nowhere in the foregoing matter. Since they have nothing to do with the "and how?" query with which the title to D2.7 ends, one must search elsewhere for the missing questions. Fortunately they are similar to questions raised in Jesuit logic notes at this point, particularly in Lorinus [LL495] and in Vallius [VL2: 255], from which the inserted title has been reconstructed; for details, see Lat. Ed. 249–252. The second piece of evidence is Galileo's difficulty in numbering the questions in this second disputation of the treatise on demonstration. A similar problem with numbering has already been seen in the third disputation of the treatise on foreknowledge, where Galileo went directly from F3.2 to F3.4, omitting F3.3 in his enumeration (see F3.4 n. 1). Here the problem is more serious, for the numbering of the questions between D2.4 and D2.10 is left unresolved by Galileo. He clearly did leave room for five questions, but only four of these are formulated explicitly. Possibly he was thrown off by D2.6 being a twofold question and thus did not search further into the difficulty. Actually the missing question was glossed over in what he may have thought was a fuller answer to D2.7, for it is clearly answered here in D2.8.

² *modes of speaking essentially*: Lat. *modus dicendi per se*, again rendering *per se* as essentially, as explained in D2.7.2 n. 10.

³ *text 23*: that is, in Bk. 5, ch. 18, 1022a15-37.

⁴ *in and by itself*: Lat. *per se*, usually translated "essentially" in this question and here having the same connotation but rendered differently to make the thought more explicit.

⁵ *ultimate genus...ultimate differentia*: that is, the genus and the differentia closest to the atomic species, as opposed to a genus and a differentia that are remote from it.

⁶ *You say, on behalf of Scotus*: all of the objections in this paragraph are based on Scotistic teaching, generally as formulated by Antonius Trombetta; the replies are the standard Thomistic responses to Scotus's emendations. Fuller details are furnished by Lorinus [LL502] and Vallius [VL2: 257].

⁷ *[identical]*: omitted by Galileo and here supplied for sense.

⁸ *[reason]*: again omitted by Galileo and supplied for sense.

⁹ *more immediately*: Lat. *magis secundum se*.

¹⁰ *text 59*: that is, in Bk. 7, ch. 17, 1041a6- 32.

¹¹ *"proper"*: Lat. *proprius*, also translatable as "property."

¹² *to every instance and solely and at all times*: Lat. *omni, soli, et semper*.

¹³ *"Man becomes gray haired"*: note that here and in the first sentence of this paragraph Galileo puts this proposition in the second mode of speaking essentially, while admitting that it belongs there "less perfectly" than the other examples. Lorinus, on the other hand, rejects it from this mode because of its temporally restricted character [LL507].

¹⁴ *an extrinsic cause*: on Galileo's use of this expression, see D1.2.5 n. 12.

¹⁵ *but the subject also*: in his revised version Vallius offers a fuller explanation of why the subject must be included: "...the moon is placed in the definition of the eclipse because, although the eclipse depends on a cause that is extrinsic to the subject, nonetheless there is some causality in the moon itself, namely, because the moon does not have its light from itself, for if there were light in the moon from within as there is in the sun or any other star, the moon would not properly be eclipsed but would always retain its own intrinsic light" [VL2: 262].

¹⁶ *I reply in the affirmative*: Galileo added the word "first" after "I reply," suggesting that he may have had more than one reason but then failed to include any additional arguments in his response.

¹⁷ *indirect and unnatural*: Lat. *indirectae et innaturales*, that is, opposed to our natural way of speaking.

¹⁸ *[are]*: omitted by Galileo and supplied for sense.

¹⁹ *first substances...second as well*: "first substance" refers to an individually existing substance with all its attributes and accidental modifications; "second substance" refers to substance taken as a category and as a universal, precision being made from its individual existence. Apparently this conclusion was disputed among commentators, with Philoponus,

Thomas Aquinas, Walter Burley, and Paul of Venice holding for the affirmative and Themistius and Grosseteste for the negative.

[20] *Grosseteste*: Lat. *Lincolniensis* or *Linconiensis*, the scholastic way of referring to Robert Grosseteste, Bishop of Lincoln.

[21] *this {third} mode*: in place of "third" Galileo wrote "second," here emended for sense.

[22] *Aristotle said*: probably a reference to text 23 of Bk. 5 of the *Metaphysics*, as in n. 1 above.

[23] *all four causes*: that is, the final, efficient, formal, and material cause. Not all Aristotelians were agreed on this point, as Lorinus points out in his commentary: "What genus of cause this mode involves is very much in doubt. Averroes thinks it contains only the efficient cause, Paul of Venice the efficient and final, but especially the final,... Giles [of Rome] these two also, but especially the efficient... Yet others expressly state that all [four] kinds of cause are included in this mode, and St. Thomas definitely favors their view" [LL510].

[24] *You ask, first*: Lorinus raises the same query but he identifies it as a notation that is directed against Cajetan and Paul of Venice: "It should be noted that a cause as it respects an effect is put in the fourth mode not only as it respects it in potency, which is Cajetan's opinion..., nor only as it respects it in act, as holds Paul of Venice, but in both ways" [LL511].

[25] *of throat-cutting*: Lat. *de iugulatione*; Aristotle cites this in Bk. 1, ch. 4 [73b15–17], as an example of an efficient cause that functions essentially, and not merely accidentally, to bring about the actual death of an animal.

[26] *a cause as it respects an effect in act is contingent*: see the related discussion in D2.7.5–6 and the accompanying notes.

[27] *also one of predicating*: there are only two modes of predicating, and these are explained in the following question, D2.9.7–11.

[28] *Apollinaris*: also known as Offredus Cremonensis, a late fifteenth-century commentator who wrote an exposition of the first book of the *Posterior Analytics*, published at Venice in 1493 and 1497 and again at Cremona in 1581.

[29] *not the two of predicating*: Galileo experienced some difficulty here, for he first wrote "two of causing," then deleted the "of causing" and wrote "of predicating" instead.

[30] *the two modes {of predicating}*: the confusion noted in the previous sentence shows up here again, for Galileo's manuscript has "two modes of causing" here, which is corrected and here shown in braces.

D2.9: What are the rules for propositions in the first and second modes? *For background, see Secs. 4.3 and 4.8 of* Galileo's Logic of Discovery and Proof, *for an application, Sec. 5.1*

[1] *[Ninth Question]*: omitted by Galileo and inserted here to maintain continuity to the following question, where he resumed numbering his questions.

[2] *the first and second modes*: that is, modes of speaking essentially, discussed at length in D2.8.

[3] *two modes of predicating*: Lat. *duo praedicandi modi*, understood in the sense of modes of predicating essentially, *modi praedicandi per se*, an expression with narrower

signification than the *modi dicendi per se* treated in the previous question, as explained in what follows.

⁴ *a natural necessity*: on the various types of necessity, see the discussion in D2.7.1 and its accompanying notes.

⁵ *a contingent subject*: that is, Peter, an individual or singular substance who comes to be and passes away. If the subject were "Man," on the other hand, the proposition would be in the second mode, as stated in D2.8.2.

⁶ *the word "is"*: a similar precision concerning the use of "is" as a logical copula or as a synonym for "exists" in its ontological sense has been made in F3.5.3 and F3.5.5 and in notes 8 and 11 of that question.

⁷ *said of every instance*: Lat. *de dicto de omni*, as in D2.7 n. 2.

⁸ *reducible*: simply by interchanging the subject and the predicate in the respective propositions. Lorinus notes that this rule is directed against Cajetan, as is the discussion of non-natural propositions in D2.8.5 [LL499].

⁹ *"A body is in place essentially"*: Lat. *corpus est per se in loco*, translating *per se* with "essentially" as heretofore.

¹⁰ *true and perfect definition of the subject*: Lat. *vera perfectaque definitio subiecti*, presumably because such a definition can only be given of a real existent, and neither a vacuum nor a chimera satisfy this requirement.

¹¹ *Aristotle proves*: that is, with the arguments he offers in Bk. 2, ch. 4, 286b10–287b22, concluding to the sphericity of the heavens, and those in Bk. 2, ch. 14, 297a8–298a20, concluding to the sphericity of the earth.

¹² *are not true and perfect demonstrations*: Lorinus offers a similar evaluation of such arguments: "I reply: these should rather be called necessary arguments than true demonstrations, or as Alfarabi says, following Averroes in comment 41, they can be spoken of as somewhat imperfect demonstrations, for in perfect demonstrations the predicate should be convertible with the subject" [LL509]. Later he adds: "There are many such demonstrations, both among physicists and among mathematicians, as when they show that the earth is round; for roundness is a more extensive term than earth but less extensive than element when that is taken in its entire scope" [LL519]. See also D2.12 n. 2.

¹³ *that is improper*: understanding property in the strict sense, in the sense of said of every instance and essential, since some earth, for example, is patently not round.

¹⁴ *two modes of predicating essentially*: Lat. *duo modi praedicandi per se*.

¹⁵ *there are five predicables*: Lat. *quinque sunt praedicabilia*. In scholastic thought the predicables are the five ways of classifying universals that are used in predicating, namely, genus, difference, species, property and accident; they differ from the ten Aristotelian categories or *praedicamenta*, which, though used in predicating, also designate modes of being as ontological categories. For the different ways the predicables are seen to predicate, see nn. 18–21 below.

¹⁶ *mode of predicating accidentally*: Lat. *modus praedicandi per accidens*.

¹⁷ *is taken from a mode of being*: Lat. *desumitur a modo essendi*.

¹⁸ *in quiddity*: Lat. *in quid*, that is, the way in which a genus is predicated of a subject, as in "Man is an animal."

¹⁹ *in quiddity of a qualitative kind*: Lat. *in quale quid*, that is, the way in which a differentia contracts a genus to a particular species, as in "Man is rational," or the way in which the species itself is predicated of a subject, as in "Man is a rational animal."

²⁰ *in quality convertibly*: Lat. *in quale convertibiliter*, the way in which a true property is predicated of a subject, as in "Man is risible."
²¹ *in quality non-convertibly*: Lat. *in quale non-convertibiliter*, the way in which an accident is predicated of a subject, as in "Man is white."
²² *of causing*: see D2.8.1 and D2.8.7.

D2.10: What modes serve the purposes of demonstration? *For background, see Secs. 4.5 and 4.8 of* Galileo's Logic of Discovery and Proof, *for an application, Sec. 3.4c n. 21*

¹ *Tenth Question*: with this question Galileo resumes his numbering of the questions in the disputation, which he left off after the fourth.
² *the modes*: that is, the modes of speaking essentially, as explained in D2.8.
³ *serve the purposes of demonstration*: Lat. *demonstrationi inservientes*, literally, "serving demonstration," understood in the sense of serving the needs or purposes of demonstration.
⁴ *Albert*: Here Lorinus gives a fuller citation of authorities: "First opinion: Themistius, Philoponus, Averroes, Albert, if we believe Pamphilus Montius in his scholia on Grosseteste, and Grosseteste and Pamphilus themselves, hold that the first two modes enter into a demonstration. Second: Soto, also indicated by St. Thomas, lect. 3 of the first book of the *Posterior Analytics,* holds that the fourth also enters. Third: and more correctly, Albert, tract. 2, chap. 11, and the Parisian Doctors, according to Paul of Venice, hold that all four modes enter. Fourth: St. Thomas, lect. 10, holds that the second and the fourth enter, and Burley indicates the same, and Dominic of Flanders, q. 15, a. 2, interprets this as applying to the conclusion. Fifth: Cajetan, when considering the conclusion, holds that the second and the fourth modes can be in it, but considering the whole demonstration, all four modes. Sixth: the opinion of certain others, according to Giles [of Rome], is that only the second mode pertains. Seventh: Alfarabi, as is gathered from Averroes, comment 31, holds that only the fourth mode pertains. Eighth: Giles attempts to reconcile all of these opinions, but himself inclines more to the second and the sixth" [LL512].
⁵ *Note*: the point of the threefold distinction introduced in this brief notation, which is lacking in the parallel expositions, is not explained by Galileo. Judging from the terminology employed in the various conclusions, however, one may surmise that the first, "with respect to the terms," prepares the way for the second conclusion; the second, "with respect to the cause," for the third conclusion; and the third, "with respect to predication," for the first conclusion.
⁶ *that make use of essential predication*: Lat. *habentibus praedicationem per se*, another way of saying "predicating essentially."
⁷ *of existing in and by itself*: Lat. *existendi per se*, as in D2.8 n. 4.
⁸ *composed of propositions*: that is, of complexes of subject and predicate and not of non-complexes such as substances, to which the third mode properly applies; see D2.8.6.
⁹ *formal cause...efficient cause*: for a discussion of these types of causality and how they serve in a demonstration, see the excerpt from Vallius's *Logica* translated in D2.2.4 n. 9.
¹⁰ *include only extrinsic causes*: Lorinus identifies this as an Averroist objection: "Averroists say that only extrinsic causes pertain to the fourth mode, and they do not enter into a demonstration unless perhaps they are convertible; Balduinus disagrees with them,

comment 35. But since they are convertible, they bring to the demonstration the aspect (*ratio*) of a formal cause, Zabarella, Bk. 2, *On necessary propositions*, ch. 14" [LL516].

[11] *they also contain intrinsic causes*: an example that illustrates this is the demonstration of a lunar eclipse through the interposition of the earth between the sun and the moon. In this case the earth is an extrinsic cause of the eclipse, but the moon also exercises causality as an intrinsic cause, as explained in D2.8.4 n. 14. See D1.2.5 n. 12; also Sec. 3.4c n. 21 of *Galileo's Logic of Discovery and Proof*.

[12] *text 10*: Bk. 1, ch. 4, 73b17–26.

[13] *the beginning of the chapter*: i.e., at text 7 of ch. 4, 73a24.

[14] *text 9*: Bk. 1, ch. 4, 73a35–b17.

[15] *text 5*: Bk. 1, ch. 2, 71b10–72b4.

D2.11: What is a universal predication? *For background, see Secs. 4.3 and 4.8 of* Galileo's Logic of Discovery and Proof, *for a relevant discussion, Sec. 2.7a*

[1] *a universal predication*: Lat. *praedicatum universale*, literally, "universal predicate," but taken in the sense of a predicate used to form a proposition, so as to cohere with the second part of the query.

[2] *said of every instance, essential, and commensurate*: Lat. *de omni, per se, et secundum quod ipsum*, that is, at text 11 of ch. 4, 73b27. See D2.7 n. 2 and D2.7.3 nn. 13–14.

[3] *as we have explained above*: in D2.7.3.

[4] *text 11*: Bk. 1, ch. 4, at 73b30.

[5] *what belongs to an adequate subject*: Lat. *convenit adaequato subiecto*, with "adequate" here meaning a subject that is adequated to, or has the same extension as, the predicate. See D2.7.4 n. 16; also Sec. 4.3a of *Galileo's Logic of Discovery and Proof*.

[6] *universal or commensurate*: Lat. *universale seu secundum quod ipsum*.

[7] *belongs to a thing precisely as such*: Lat. *convenit rei ut talis est*.

[8] *text 11*: at 73b27–28.

[9] *text 14*: Bk. 1, ch. 5, 74a33–b4.

[10] *texts 11 and 19*: Bk. 1, ch. 4, 73b27–74a4, and ch. 6, 75a29–37.

[11] *of a primary subject*: Lat. *de subiecto primo*, that is, a subject that is first and adequate, as in D2.7.4.

[12] *many authors have many opinions*: Lat. *multi multa dicunt*. In his *Logica* Vallius lists four different schools of thought on the problem, as follows. The first is attributed by Averroes to Alfarabi and Avempace, holding that there are three kinds of universal predicate (cf. D2.12.3); the second holds for two kinds, attributed to Albert and St. Thomas and including also Themistius; the third holds for one universal according to form (*secundum formam*) and another according to matter (*secundum materiam*), attributed to Paul of Venice, Apollinaris, and other Latins; and the fourth again holds for three types, though different from those of the first opinion, attributed to Averroes [VL2: 277].

[13] *there are many and various grades*: Lat. *varios et multiplices esse gradus*. *Gradus* may also be translated as "step," "stage," or "degree," but in this case "grade" seems the more appropriate.

[14] *and so on*: Vallius adds only risibility to this enumeration [VL2: 277–278], but others

could be added, such as morality, sociability, educability, responsibility, imputability, etc.

[15] *precisely as such*: Lat. *secundum quod talis est.*

[16] *in the first mode*: that is, in the first mode of speaking essentially, as in D2.8.2.

[17] *universal propositions, or those that are commensurate*: Lat. *propositiones universales, seu secundum quod ipsum*, here applying the notion of "commensurate universal" not only to the predicate but to the proposition of which it is a part.

[18] *belong to the object precisely as such*: Lat. *conveniunt rei prout talis est.*

[19] *texts 11 and 14*: Bk. 1, ch. 4, 73b27–29, and ch. 5, 74a38–39.

D2.12: Must perfect demonstrations be made from propositions that are essential, universal, and proper? *For background, see Secs. 3.4, 4.2, 4.6, and 4.8 of* Galileo's Logic of Discovery and Proof, *for an application, Sec. 6.1*

[1] *Last Question*: Lat. *Quaestio ultima*, the twelfth question in this disputation.

[2] *perfect demonstration*: Lat. *demonstratio perfecta.*

[3] *made from propositions that are essential, universal, and proper*: Lat. *ex propositionibus per se, universalibus, et propriis*, with *universalibus* being understood in the sense of commensurately universal propositions, as explained in D2.11.

[4] *accidental kinds of being*: Lat. *entia per accidens.*

[5] *universals*: understood in the sense of commensurate universals.

[6] *from proper propositions that are not primary, etc.*: Lat. *ex propriis non primis...ex primis non propriis...neque ex primis neque ex propriis.*

[7] *composed {of} propositions that are merely accidental*: Lat. *constare ex propositionibus per accidens*, but in place of *ex* here Galileo unaccountably wrote *et*, here corrected and thus enclosed in braces.

[8] *from a true and proper cause*: Lat. *per veram propriamque causam.*

[9] *by induction*: Lat. *inductione*, that is, by going through the various kinds of cause – intrinsic (formal and material) as well as extrinsic (final and efficient) – and verifying that explanations through all of them involve essential propositions. See Sec. 4.6b of *Galileo's Logic of Discovery and Proof.*

[10] *beyond the intention of the agent, or happening rarely*: Lat. *praeter intentionem agentis, raro contingere*. Usually the first type will involve impediments or unforeseen events, and these "can fall under a science" provided the appropriate suppositions are made, particularly with demonstrations formulated *ex suppositione finis*. Rare events such as an eclipse or a rainbow or a conjunction of planets can be handled similarly, on the supposition of the appropriate positions, times, and causal factors being otherwise present. See F3.1.7 n. 22, D2.2.5 n. 10, D2.4.2 n. 8, and D2.7.7 n. 22; also Sec. 4.2 of *Galileo's Logic of Discovery and Proof.*

[11] *the demonstration by which Aristotle proves that the earth is at rest is not a true demonstration*: that is, the argument elaborated in *De caelo*, Bk. 2, ch. 14, 296a24–297a8. One can only wonder whether this early statement by Galileo could have had any influence on his many later attempts to demonstrate the earth's motion.

[12] *accidents*: Lat. *accidentia*, meaning by this praedicamental accidents in the nine categories of being that may have some necessary connection with their subjects, and not

predicable accidents that lack such a connection; see D2.9.7 nn. 14–16, D2.9.9 nn. 18–21.

¹³ *universal*: again understanding this as commensurately universal.

¹⁴ *text 11*: Bk. 1, ch. 4, 73b27–74a4.

¹⁵ *in the following demonstration*: the point of the example is that, while it is essential to the various species of animal to sense and to move locally, the peculiar ways in which they sense and so move are proper to each species, and thus one cannot demonstrate commensurately universal properties of the appropriate kind by repeatedly invoking the same middle term.

¹⁶ *and this on the part of the subject and on the part of the predicate*: that is, the subject must be the unique subject in which the predicate may be found, and the predicate must be a strict property that can be predicated uniquely of that subject.

¹⁷ *mathematical demonstrations...are not perfect*: In his *Logica* Vallius elaborates on this point, noting that mathematicians do not generally invoke true causes, and citing Euclid's famous proof of the sum of the interior angles of a triangle, which he says is made through the use of a construction and not directly from the nature of a triangle. He also mentions their frequent use of demonstrations to the impossible, and the fact that they use common principles from which they deduce many conclusions without being able to furnish a proper explanation for each [VL2: 285–286]. See D2.5.8 n. 14 and D2.6.5 n. 15.

¹⁸ *which [is] the cause of uncertainty*: Lat. *quae causa incertitudinis est*. In writing this Galileo unaccountable left out the "is" (*est*), here shown in brackets. Matter is the cause of uncertainty because it frequently is the source of defects or imperfections that impede the operation of natural causes. See F4.1.9 and D2.6.6 n. 17.

¹⁹ *most powerful demonstration proceeds...from the definition, etc.*: on the close connection between definition and most powerful demonstration, note the many references to their interrelationships in D1.2, e.g., in paragraphs [18], [26], and [29].

²⁰ *not universal*: again in the sense of not being commensurately universal.

²¹ *of individuals*: because they are contingent subjects; see D2.9.1 n. 5.

²² *God is singular*: note the analysis of the comparable statement, "God is unchangeable," in D2.7.2 (at n. 12).

D3: *On the Species of Demonstration*

D3.1: How many species of demonstration are there? *For background, see Secs. 4.4, 4.6, and 4.9 of* Galileo's Logic of Discovery and Proof, *for an application, Sec. 6.6b*

¹ *On the Species of Demonstration*: Lat. *De speciebus demonstrationis*, understanding species in the strict sense of kinds or types contained under the genus of demonstration.

² *is said to be topical reasoning by Aristotle*: probably a reference to Bk. 1, ch. 1, of the *Topics*, 100a30, where Aristotle says that reasoning based on common principles (*endoxa*) is dialectical or topical. Lorinus's parallel statement here is a bit clearer than Galileo's and translates as follows: "Demonstration is sometimes taken broadly for all necessary reasoning, as it is probably taken in the first book of the *Topics*, chapter 1, when he [Aristotle] divides syllogisms into necessary, topical, and sophistic. At other times it is taken more strictly for reasoning made from necessary and proper premises, for which reason in the first book of the *Posterior Analytics*, text 30 [i.e., ch. 13, 78a23–25], reasoning based on

common principles, although sometimes necessary, is said to be logical reasoning... It is said to be logical or even topical reasoning when compared with demonstration that proceeds from necessary and proper principles, whereas it is said to be demonstration when compared with a syllogism made only from probable or apparent principles" [LL524].

³ *taken in the second way*: probably a reference to the second of the two ways noted in paragraph [1], that is, as a reasoning process based on necessary and proper principles, and so, as Lorinus puts it, as it is "a perfect instrument of scientific knowing" [LL524].

⁴ *Averroes*: as explained in paragraph [7] below.

⁵ *Franciscus [Neritonensis]*: Galileo seems to have had difficulty reading his exemplar here, for he left a space after *Franciscus*, apparently to be filled in later. Lorinus gives the correct reading: "...the opinion of Franciscus Neritonensis, as Balduinus refers to it in q. 70, that demonstration of the reasoned fact is not true demonstration but only demonstration of the fact" [LL532].

⁶ *cited by Averroes in many places*: Lorinus again fills in the missing information, "in comments 55 and 59, and in question 15" [LL532].

⁷ *in the second book of the Prior Analytics*: that is, in chapter 16, 64b32–33.

⁸ *we beg the question*: Lat. *nos petere principium*; see F3.5.3 n. 7.

⁹ *there is no circular demonstrative regress*: Lat. *non dari regressum demonstrativum et circularem*; that is, in text 6, 72b25–73a21; the problem is taken up in detail in D3.3.

¹⁰ *more known with respect to nature...with respect to us*: see D2.6.1.

¹¹ *propositions that are indirect and unnatural*: see D2.8.5 n. 16.

¹² *"Every risible [being] is a rational animal"*: Lat. *omne risibile est animal rationale*. The word "being," shown in brackets, is inserted to make sense in English; when omitted, as it is in the Latin, the sentence becomes literally "Every risible is a rational animal," an unnatural way of speaking.

¹³ *it begs the question*: Lat. *petitur principium*; again see F3.5.3 n. 7.

¹⁴ *must be known essentially and not through the senses*: Lat. *per se, non per sensum, cognoscenda est*.

¹⁵ *of others*: In his *Logica* Lorinus also names Antonius Scotius, Tomitanus, Alexander Piccolomineus, Marcus Antonius Ermia, Melioratus, and "almost all Averroists" [LL524].

¹⁶ *from texts 5 to 30*: that is, chapter 2 through chapter 12.

¹⁷ *from text 30 onwards*: that is, chapter 13 through chapter 34.

¹⁸ *text 42*: that is, in chapter 27.

¹⁹ *text 7*: that is, in Bk. 2, ch. 7.

²⁰ *the first book of the Topics, chapter 1*: at 100a25–101a4.

²¹ *"in knowing."..."in being"*: Lat. *in cognoscendo...in essendo*.

²² *its qualities are reducible to a complex existent*: Lat. *qualis sit reducitur ad esse complexum*; for example, man's rationality or his risibility are reducible to the propositions "Man is rational" or "Man is risible," both of which are complex entities because composed of a subject and a predicate.

²³ *of Thomas, etc.*: Lorinus expands the list to include Albert the Great, Robert Grosseteste, and Apollinaris among the Latins, and Alexander of Aphrodisias (according to Augustinus Niphus) among the Greeks [LL531]. Vallius adds to these Giles of Rome, John of Jandun, and Zabarella [VL2: 314].

²⁴ *text 30*: that is, in chapter 13 of Book 1.

²⁵ *Alexander*: Here Vallius adds Themistius and Philoponus on the first book of the

Posterior Analytics, chapter 7 [VL2: 306].

[26] *that serve the needs of science*: Lat. *quae scientiae deserviunt*, that is, being based on singulars and not on propositions that are universal, it lacks what is required for knowing scientifically.

[27] *by the very way it works*: Lat. *quantum in se est*, literally, "as much as it is in itself," an expression used by Descartes and Newton to characterize a body's inertial tendency to persevere in straight-line motion, in their context translated "as much as in it lies."

[28] *a thing is related to being as it is to knowing*: Lat. *res ita se habeat ad esse quoad cognosci*, a realist principle frequently invoked in these notes; see F3.1.4, F4.2.3, D1.1.11, D2.2.7, D2.2.9, and D2.5.12.

[29] *in the sixth book of the Topics*: in chapter 4, 141a26–b2.

[30] *from true causes...intrinsic or extrinsic*: Lat. *ex veris causis ...intrinsecae vel extrinsecae*. Vallius here replies to an objection against the propriety of using extrinsic causes with the following comment: "The most this argument establishes is that this type of demonstration [i.e., the type made from extrinsic causes] is not as perfect as one that demonstrates an intrinsic property that depends on the nature and quiddity of the subject as on an adequate and total cause, and this we readily concede. For these extrinsic properties depend both on the nature of the subject and on an extrinsic cause, and therefore they are not always found with the subject as are intrinsic properties, but only when the extrinsic cause is present" [VL2: 333].

[31] *in the first book of the Posterior Analytics*: probably in chapter 11, where demonstration through impossibility is mentioned at 77a23–24.

[32] *either in the premises or prior to the premises*: Lat. *vel in praemissis vel ante illas*. How this expression is to be understood is the burden of the question on the demonstrative regress, D3.3.

[33] *we beg the question*: Lat. *peti principium*; see F3.5.3 n. 7.

[34] *a topical or probable syllogism*: Lat. *syllogismus topicus vel probabilis*; see paragraph [1] and n. 2 above.

[35] *against the ancients*: Lat. *contra veteres*; their identity is not known, but possibly they were followers of Xenocrates.

[36] *we attain perfect science by arguing demonstratively in a circle*: Lat. *per circulum demonstrativum acquiri perfectam scientiam a nobis*. The emphasis here should be on the word "perfect," as seen in the following note, for Galileo is willing to admit that some circularity is involved in the use of demonstration of the fact, which is "imperfect" when compared with demonstration of the reasoned fact, as admitted in paragraph [16] below.

[37] *Supply "perfect."..and one begs the question*: Lat. *Subintellige perfectum, et peteretur primum principium*. That is, if the expression "perfect science" were not in the previous sentence there would be no begging of the question; when it is put there, there is.

[38] *by experience...by induction...by the light of the intellect*: Lat. *experientia... inductione... lumine intellectus*; here following the commentary of St. Thomas Aquinas on the second book of the *Posterior Analytics*, chapter 19. Note that, despite his earlier references to the limitations of inductive reasoning in paragraphs [6] and [10], Galileo is here maintaining that the human intellect has the capability of arriving at true and certain propositions by the process of induction. This is why he lists the "intellectual light" (*lumen intellectuale*) among the instruments of knowing in D1.2.2. See Secs. 2.1, 2.2, and 4.6 of *Galileo's Logic of Discovery and Proof*.

[39] *otherwise nature would have looked unkindly...proper conditions and necessary properties*: Lat. *alias natura male huic universalitati prospexisset, quippe quae rebus suas conditiones et proprietates necessarias non dedisset*. The sentence is difficult to translate, but the sense is that nature would make herself unintelligible to man if the human intellect were not able to discern necessary connections and understand the conditions under which they obtain in the order of nature.

[40] *from text 5 to 30*: Bk. 1, chs. 2 to 12.

[41] *understood [of] most powerful demonstration*: in writing this Galileo left out the "of" (*de*), here supplied for sense.

[42] *in the second book of the Posterior Analytics*: in chapter 11, 94a20–95a9.

D3.2: How are demonstrations of the reasoned fact and of the fact similar and dissimilar? *For background, see Secs. 4.4 and 4.9 of* Galileo's Logic of Discovery and Proof, *for an application, Sec. 5.1*

[1] *How...similar and dissimilar*: Lat. *In quo conveniant et differant*.

[2] *on the division of the latter*: Lat. *de huius divisione*.

[3] *analogically the same*: Lat. *analogice inter se convenire*; on analogical similarity, see D1.2.29 n. 30 and D2.4.5 n. 13. Vallius has a fuller explanation in his *Logica*: "...demonstration of the fact, from an effect, is less perfect than demonstraton of the reasoned fact, and thus it is not demonstration as properly as is demonstration of the reasoned fact; rather the formality (*ratio*) of demonstration is found in them analogically and not univocally, and for this reason demonstration in general (*demonstratio in communi*) is not truly and properly a genus for them under the aspect being considered" [VL2: 307].

[4] *text 30*: ch. 13.

[5] *demonstration "of a sign"*: Lat. *demonstratio signi*.

[6] *text 19*: ch. 6, at 75a33.

[7] *demonstration "of the fact" or "by which"*: Lat. *demonstratio quia vel quo*.

[8] *demonstration "of evidence"*: Lat. *demonstratio evidentiae*.

[9] *demonstration "of existence"*: Lat. *demonstratio existentiae*.

[10] *demonstratio "from an effect" or "a posteriori"*: Lat. *demonstratio ab effectu seu a posteriori*.

[11] *"conjectural"*: Lat. *coniecturalis*.

[12] *"a wolf does not reason"*: Galileo's example here seems to be defective, for many people would regard the ability to reason as the proximate cause (and not a remote cause) of a sense of wonder. Yet his example is consistent with the syllogism given in Vallius-Carbone's *Additamenta* [1vb–2ra], where, as is analyzed in Sec. 2.7a of *Galileo's Logic of Discovery and Proof*, they introduce two middle terms between being rational (i.e., being able "to reason") and being capable of wonder (i.e., having "a sense of wonder"), namely, being "capable of discourse" and being "able to recognize an effect before a cause." In this context Aristotle's own example of a remote cause would have been better: "A wall is not an animal; therefore it does not breathe" (78b15–27). Since not all animals breathe, being an animal is only a remote cause of breathing – having lungs, on the other hand, is the proximate cause. Galileo himself makes use of this example in D3.3.14.

[13] *there is smoke, therefore fire*: apparently Galileo takes smoke to be a proper effect or a

necessary sign of fire and so convertible with it. Lorinus rejects that particular connection: "From smoke one cannot infer the existence of fire by a true demonstration, for not all smoke is effected by fire" [LL552].

[14] *heating occurs, therefore there is fire*: it is not clear here whether Galileo is proposing this as a true demonstration of the fact, or simply as an argument involving non-convertible terms; see the preceding note.

[15] *protomatter*: Lat. *materia prima*, in the *Physics*, Bk. 1. Vallius gives the same example in his *Logica* [VL2: 348] and attributes its identification to Zabarella.

[16] *a first mover*: Lat. *primus motor*, in the *Physics*, Bk. 8 – also cited by Vallius in this place and attributed to Zabarella. See also D2.6.2.

[17] *the [element] fire*: Lat. *ignem esse*, probably in *On Generation and Corruption*, Bk. 2, chs. 1–3.

[18] *are most useful in the sciences*: Lat. *in scientiis esse utilissimas*, meaning by this the natural sciences, those concerned with the world of nature, where things are most known with respect to nature and not most known to us. See D3.3.13.

D3.3: Is there a demonstrative regress? *For background, see Secs. 2.8, 4.4, and 4.9 of* Galileo's Logic of Discovery and Proof, *for applications, Secs. 5.1, 5.2, 6.2, 6.5b, and 6.7a*

[1] *demonstrative regress*: Lat. *regressus demonstrativus*.

[2] *ancient philosophers, referred to by Aristotle*: in *Posterior Analytics*, Bk.1, ch. 3, 72b25–73a21, as already cited in D3.1.5 n. 9. See also D3.1.15 n. 35.

[3] *a perfect circle*: Lat. *perfectus circulus*, as opposed to the imperfect circle endorsed in paragraph [5] below. See also D3.1.15 n. 36.

[4] *took demonstration of the fact from its middle position*: Lat. *de medio substulerunt demonstrationem quia*; an expression whose sense is obscure but was probably intended to mean that they took demonstration of the fact from its mediating role in supplying terms from which the demonstration of the reasoned fact could be formulated and thus made the regress itself impossible.

[5] *one...begged the question*: Lat. *petitur principium*. See D3.1.5–6 and D3.1.15; also D3.2.5–6.

[6] *[Ugo Senensis]*: this name was omitted by Galileo and a space left for it, presumably to be filled in later. The name is correctly given in Lorinus's notes of his lectures on logic given at the Collegio Romano in 1584 and again in the revised version of Vallius's course [Lat. Ed. 292; VL2: 346].

[7] *Neritonensis*: written incorrectly by Galileo as *Eritonensis*; possibly he misread the "N" as an "H" and dropped it, as was his custom when copying Latin words that begin with "h."

[8] *second progression*: Lat. *secundus progressus*, literally "second progress" but translated as "second progression" to agree with Galileo's use of *duae progressiones* in paragraph [14] below.

[9] *in the second book of the Prior Analytics*: ch. 16, 64b32–22, as also cited at D3.1.5.

[10] *Neritonensis's*: again written without the "N."

[11] *Aristotle's*: Lorinus gives a much fuller citation of authorities for this position, including Apollinaris, Albertus [Magnus], Caietanus, Aegidius [Romanus], Paulus Venetus, Themistius, Averroes, Dominicus de Flandria, and Zabarella [LL554].

[12] *the proving part and the part proved*: Lat. *id quod probat et id quod probatur.*

[13] *under the formal relationship of cause and effect*: Lat. *sub relatione formali causae et effectus.*

[14] *as they are disparate things*: Lat. *quatenus sunt res disparatae.*

[15] *as the cause is necessarily connected with the effect*: Lat. *quatenus causa necessario est connexa cum effectu.*

[16] *correlatives are so interdependent that one cannot be more known than the other*: Lat. *relativa ita se habent inter se, ut unum non sit notius altero.* That cause and effect are correlatives has previously been stated in F2.3.1; the principle itself has already been invoked implicitly in D1.2.22.

[17] *there is no circularity*: Lat. *non dari circulum.*

[18] *a progression of reasoning in demonstration*: Lat. *progressus rationis in demonstratione*, taking *progressus* in the sense of *processus*, that is, a reasoning procedure that could include both the forward and the backward (or regressive) motions found in the demonstrative regress. This is the sense in which the term is used by Zabarella and Lorinus; see Lat. Ed. 294–296.

[19] *for the more perfect development of the sciences*: Lat. *propter perfectiorem scientiarum inventus*, taking *inventus* to mean development by way of discovery (as in the *via inventionis*, thus as opposed to the *via doctrinae*, by way of doctrine or teaching). See Sec. 2.8 of *Galileo's Logic of Discovery and Proof.*

[20] *in the {third} way*: reading Galileo's *secundo* as *tertio*, another mental lapse on his part, and thus enclosed in braces.

[21] *not considered under the same formality and not to the same term*: Lat. *non eadem ratione et neque ad idem.*

[22] *[its] effect*: the word for "its" (*eius*) is missing in Galileo's manuscript but a space was left for it to be filled in later; the *eius* here is a conjectured reading.

[23] *[A regress of this kind is made when]*: these words are likewise missing in Galileo's manuscript, though space has been left for them; the reading is conjectured, as in the preceding note.

[24] *perfect circularity*: Lat. *circulus perfectus.*

[25] *there is no progression to the same numerical term*: Lat. *non fit progressus ad idem numero.*

[26] *in the second progression*: Lat. *in secundo progressu.*

[27] *for the perfecting of all the sciences*: Lat. *perfectioni omnium scientiarum*, understanding "perfecting" as in n. 19 above.

[28] *the conditions for the demonstrative regress*: Lat. *conditiones regressus demonstrativi.* For a detailed comparison of the conditions given by Galileo with those enumerated by Zabarella, Lorinus, and Vallius, see Lat. Ed. 298–302.

[29] *[six in number. First]*: these words are missing in Galileo's manuscript but a space was left for them, apparently to be filled in later; the reading supplied here is conjectured.

[30] *that there be two progressions of demonstration in it*: Lat. *ut in illo fiant duo progressiones demonstrationis.*

[31] *as we have explained*: i.e., in paragraph [12] above.

[32] *wait until we come to have formal knowledge of the cause we first knew only materially*: Lat. *expectemus donec causa, quam cognoscimus materialiter, formaliter cognoscamus.* Galileo's opposition here between *materialiter* and *formaliter* parallels Zabarella's

opposition between *confuse* and *distincte* in similar contexts. For the textual evidence that supports a gradual evolution of this terminology from Zabarella to Galileo *via* the logic notes of Lorinus and Vallius, see Lat. Ed. 299–301; also Sec. 4.9c of *Galileo's Logic of Discovery and Proof*. Either terminology may be illustrated using as an example the unraveling of a fictional murder mystery. Not infrequently the murderer comes to be known early on, say, from the circumstances of the plot, but he is not known at the time precisely as the murderer. In Galileo's terms he at first is known only *materialiter*, i.e., as a person, but not as the killer. Then, as the plot unfolds, he comes to be recognized as the one who actually committed the crime, and therefore *formaliter* as the murderer. (In Zabarella's terms he at first is known only "confusedly," and then later "distinctly," as the perpetrator.) The transition from one stage to the other usually takes time, during which one considers various possibilities, then excludes those less likely, and so on, all of which requires what Zabarella calls the work of the intellect (*negotiatio intellectus* or *labor mentis*). See Sec. 4.9b of *Galileo's Logic of Discovery and Proof*.

³³ *that we may have formal knowledge of the cause*: Lat. *ut a nobis cognoscatur causa formalis*, taking Galileo's *formalis* here to be the equivalent of his *formaliter* in the previous note.

³⁴ *one has formal knowledge of the cause*: Lat. *qui cognoscit causam formalem*, again taking *formalem* here as equivalent to *formaliter*.

³⁵ *to material knowledge of the cause...from formal knowledge of the cause*: Lat. *ad causam materialem...a causa formaliter cognita*, taking *materialem* here as equivalent to *materialiter cognitam*, paralleling the *formaliter cognita* in the second part of the expression.

³⁶ {*progression*}: Galileo wrote *effectum* here rather than the *progressum* that is required for sense; the emended term is thus shown in braces.

BIOGRAPHICAL AND BIBLIOGRAPHICAL
REGISTER

The bibliography here presented is restricted to the source materials on which Galileo's logical treatises are based, either proximately or remotely. For the convenience of the reader unacquainted with Greek, medieval, and Renaissance authors, brief biographical sketches are also included. With regard to the editions cited, generally these are works published prior to Galileo's writing of MS 27. Those seeking a listing of more recent works relating to this manuscript should consult that provided in our companion volume, Galileo's Logic of Discovery and Proof.

Achillini, Alessandro. Italian Renaissance philosopher, b. Bologna, 1463, d. there, 1512; studied philosophy and medicine at the University of Bologna, later served there as a professor. *Opera omnia in unum collecta...cum annotationibus Pamphili Montii*, Venice 1545.

Albert the Great, St. Dominican philosopher and theologian, b. Lauingen *c.* 1200, d. Cologne 1280; taught at Paris and Cologne, where he numbered Thomas Aquinas among his students. *Logica*, Venice 1494, 1506, 1560, Cologne 1504; *Opera omnia*, Lyons 1651, Paris 1890–1899, Cologne 1951.

Alexander of Aphoridisias. Peripatetic philosopher of the second or third century A.D. whose commentaries on Aristotle's logical works have survived in part. *Opera omnia*, Venice 1551–1553.

Alfarabi. Arab philosopher, b. Wasij *c.* 870, d. Damascus 950; studied in Baghdad and Constantinople, commented on Aristotle's *Perihermenias*, served as a principal expositor of Platonic and Aristotelian thought in Islam, drawing mainly on the works of Alexander of Aphrodisias and Themistius.

Ammonius Hermaeus. Platonic philosopher, d. Alexandria *c.* 517–526, attempted to harmonize the teachings of Plato and Aristotle. *Aristotelis liber de interpretatione, Ammonio Hermea interprete*, Venice 1543; *In praedicamentis Aristotelis*, Venice 1550.

Antonius Andreas. Spanish Franciscan, philosopher and theologian, b. Saragossa *c.* 1280, d. 1320; studied under Duns Scotus, whose teachings he energetically propagated. *Scriptum in artem veterem Aristotelis*, Bologna 1481, Venice 1508, 1517.

Apollinaris [Offredus Cremonensis]. Italian logician and philosopher, b.Cremona, fl. second half of the fifteenth century; known principally for his controversy with Peter of Mantua over the first and last instant. *Expositio in primum librum Posteriorum Aristotelis*, Venice 1493, 1497; Cremona 1581; *In librum Aristotelis de anima*, Milan 1474; *De primo et ultimo instanti*, Collensi 1478.

Avempace. Arab philosopher who flourished in Spain, b. Saragossa, d. Fez 1138; wrote treatises on logic, mathematics, and medicine, commented on Aristotle's *Physica, De generatione et corruptione, Meteorologica, De generatione animalium*, and *De partibus animalium*.

Averroes. Moslem philosopher, b. Cordova 1126, d. Morocco 1198; commented on the

216

works of Aristotle with such ability that he was widely known as "the Commentator." *Opera Aristotelis cum Averrois commentariis*, Venice 1550–1552, 1562–1574, etc.

Avicenna. Arab philosopher, b. Afshana 980, d. Hamadhan 1037, student and interpreter of Aristotle. *Opera philosophica*, Venice 1508.

Balduino, Girolamo. Italian philosopher who flourished in the mid-sixteenth century, taught logic at the University of Padua in 1528. *Expositio in librum primum Posteriorum Aristotelis*, Venice 1563; *Varii generis in logica quaesita*, Venice 1609.

Baliani, Giovanni Battista. Italian mechanician and correspondent of Galileo, b. Genoa 1582, d. there 1666. *De motu naturali gravium solidorum*, Genoa 1638; *De motu naturali gravium solidorum et liquidorum*, Genoa 1646.

Barberini, Maffeo (Urban VIII). Florentine cardinal, later pope, b. Florence 1568, d. Rome 1644; studied philosophy at the Collegio Romano, law at the University of Pisa; sided with Galileo in the dispute at Florence over bodies in water; elected pope in 1623; initiated proceedings against Galileo following publication of the *Dialogo* in 1632.

Bellarmine, Robert, St. Jesuit philosopher and theologian, b. Montepulciano 1542, d. Rome 1621; studied at the Collegio Romano, taught there and at Louvain; created cardinal in 1598, played a leading role as a defender of traditional Catholic teaching in the Counter Reformation.

Benedetti, Giovanni Battista. Mathematician and natural philosopher, b. Venice 1530, d. Turin 1590; studied mathematics with Tartaglia, wrote against Aristotle's treatment of falling bodies. *Resolutio omnium Euclidis problematum*, Venice 1553; *Demonstratio proportionum motuum localium*, Venice 1554.

Blancanus, Iosephus. Jesuit mathematician, b. Bologna 1566, d. Parma 1624; studied under Clavius at the Collegio Romano, taught at Parma, where Giovanni Battista Riccioli was in turn his student. *De natura mathematicarum scientiarum*, Bologna 1615; *Apparatus ad mathematicarum studium*, Bologna 1620.

Boethius. Philosopher and translator, b. Rome c. 480, d. near Pavia, *c.* 524; educated in Athens and Alexandria, planned to translate into Latin all of Aristotle and Plato to show the basic agreement in their teachings; his translations of the *Categories* and *Perihermenias* have survived, but that of the *Posterior Analytics* attributed to him in the Renaissance is the work of James of Venice (c. 1128). *Logica Aristotelis Boethio interprete*, Paris 1540, Venice 1543, 1549, 1551, etc.

Borro, Girolamo. Philosopher and teacher of Galileo, b. Arezzo 1512, d. Perugia 1592; taught twice at the University of Pisa, 1553–1559 and 1575–1587. *Del flusso e reflusso del mare*, Lucca 1561, Florence 1577, 1583; *De motu gravium et levium*, Florence 1575, 1576; *De peripatetico docendi atque addiscendi methodo*, Florence 1584.

Brahe, Tycho. Danish astronomer, b. Knudstrup 1546, d. Prague 1601; proposed a system of the universe that reconciled the conflicting claims of Ptolemy and Copernicus; made accurate observations of Mars, on the basis of which Kepler determined the basic laws of planetary motion in 1609 and 1619.

Buonamici, Francesco. Aristotelian philosopher and teacher of Galileo, b. Florence 1533, d. Pisa 1603; taught at the University of Pisa from 1565 to 1603; many of Galileo's invectives against the Aristotelians of the day seem directed against his teachings, preserved in his *De motu*, Florence 1591.

Buridan, John. Medieval philosopher, b. Bethune, d. after 1358; taught for about 50 years at the University of Paris, where he served as rector, 1328–1340; wrote commentaries on the *Isagoge* of Porphyry and on Aristotle's logical works. *Compendium logicae*, Venice 1499.

Burley, Walter. Philosopher and theologian, b. England c. 1275, d. c. 1345; studied at Oxford and Paris and wrote extensively on logic and natural philosophy. *Scriptum super artem veterem Porphyrii et Aristotelis*, Venice c. 1478, 1497; *Scriptum super libros Posteriorum Aristotelis*, Oxford 1517; Venice 1521.

Cajetan, Thomas de Vio. Dominican philosopher and theologian, b. Gaeta 1468, d. Rome 1534; taught at the Universities of Padua and Pavia, regarded as the foremost commentator on the writings of Thomas Aquinas. *In Praedicabilia Porphyrii, Praedicamenta, Postpraedicamenta, et libros Posteriorum Aristotelis commentaria*, Lyons 1572, 1579.

Capreolus, John. Dominican philosopher and theologian, b. Rodez c. 1380, d. there 1444; studied and taught at the University of Paris, defending the teaching of Thomas Aquinas against Scotus, Ockham, and other theologians. *Libri defensionum theologiae divi Thomae de Aquino*, Venice 1483, 1514, 1519, 1589.

Carbone, Ludovico. Logician and rhetorician, b. Costacciaro, d. Venice 1597; studied under the Jesuits at the Collegio Romano, from whom he appropriated much of his material; taught at the University of Perugia. *Introductio Toleti in dialecticam Aristotelis, additis praeludiis*, Venice 1588; *Introductio in logicam una cum catalogo auctorum qui de logica scripserunt*, Venice 1597; *Additamenta ad commentum Francisci Toleti in logicam Aristotelis: Praeludia in libros priores analyticos*; *Tractatio de syllogismo*; *De instrumentis sciendi*; *De praecognitionibus et praecognitis*, Venice 1597, 1617, 1688; *Introductio in universam philosophiam*, Venice 1599.

Carcavi, Pierre. French diplomat, b. Lyons c. 1600, d. Paris 1684; a colleague of the mathematician Pierre Fermat at Toulouse, he was later put in charge of the royal library at Paris, became one of the first members of the Academy of Sciences there in 1666.

Castelli, Benedetto. Benedictine monk and mathematician, b. Brescia 1578, d. Rome 1643; studied with Galileo at Padua, later taught at Pisa and at Rome, where his students included Bonaventura Cavalieri and Evangelista Torricelli; defended Galileo's teachings on floating bodies. *Della misura dell'acque correnti*, Rome 1628, 1639; Bologna 1660.

Cavalieri, Bonaventura. Jesuati priest and mathematician, b. Milan c. 1598, d. Bologna 1647; studied under Castelli at Pisa, who introduced him to Galileo, whom he likewise regarded as his teacher and to whom he wrote some 112 letters. *Geometria indivisibilibus continuorum nova quadam ratione promota*, Bologna 1635, 1653.

Clavius, Christopher. Jesuit mathematician, b. Bamberg 1537, d. Rome 1612; studied at Coimbra, taught mathematics at the Collegio Romano from 1565 to his death, during which time his preeminence in mathematics and astronomy was recognized throughout Europe. *In Sphaeram Ioannis de Sacro Bosco commentarius*, Rome 1570, 1581, 1585, etc.; *Euclidis Elementorum*, Rome 1589.

Commandino, Federico. Mathematician, translator, editor, b. Urbino 1509, d. there 1575; edited many Greek mathematical works, prepared Latin translations of, and commentaries on, Euclid and Archimedes. *Liber de centro gravitatis*, Bologna 1565.

Conciliator (Pietro d'Abano). Paduan philosopher and physician, b. Abano 1257, d. Padua c. 1315; his major preoccupation was the reconciliation of medicine with philosophy. *Conciliator differentiarum philosophorum et praecipue medicorum*, Mantua 1472, Venice 1520.

Doctores Parisienses. Fourteenth-century professors of the University of Paris associated with John Buridan, especially Albert of Saxony and Nicole Oresme. Galileo refers to the *Doctores Parisienses* in his MS 46 and thus knew of their teaching from the Jesuit lecture notes on which that is based.

Dominic of Flanders. Belgian Dominican who taught philosophy in Bologna, Pisa, and Florence, b. Merris *c.* 1425, d. Florence 1479. *Quaestiones 49 in primum librum Posteriorum et 20 in secundum Posteriorum,* Venice 1496; *Summa divinae philosophiae,* Venice 1499.

Eudaemon-Ioannis, Andreas. Jesuit philosopher, b. Greece, d. Rome 1625; taught logic at the Collegio Romano in the academic year 1596–1597. *Logica P. Eudemon ...Quaestiones et expositiones in universam Aristotelis logicam,* Rome, Gregorian University Archives, Fondo Curia, Cod. 511 (1597).

Eustratius. Greek commentator on Aristotle's logical writings. *In secundum librum Analyticorum Posteriorum,* Venice 1542.

Fantoni, Filippo. Camaldolese monk and mathematician, d. Volterra 1591; professor of mathematics at the University of Pisa in 1560–1567 and 1581–1589, after which Galileo succeeded him to the post; left lecture notes on geography, cosmography, and astronomy and treatises *De motu gravium et levium* and *An demonstrationes mathematicorum sint certissimae,* all preserved in Florence, Biblioteca Nazionale Centrale, MSS Conv. Soppr. B.10.480 and 481.

Ferrariensis, Franciscus Silvestri. Dominican philosopher and theologian, b. Ferrara 1474, d. Rennes 1528; taught in Dominican houses of study in Ferrara and Bologna. *Annotationes in libros Posteriorum Aristotelis et Sancti Thomae,* Venice 1517.

Foscarini, Paolo Antonio. Carmelite friar, b. Montalto in Calabria 1580, d. Montalto 1616; wrote a treatise explaining how the Copernican system could be reconciled with statements in Scripture apparently contradicting it.

Ghetaldi, Marino. Yugoslavian mathematician, b. Ragusa 1566, d. there 1626; studied under Clavius at the Collegio Romano, met Galileo in the Pinelli home at Padua before 1600, later corresponded with him. *Nonnullae propositiones de parabola,* Rome 1603; *De resolutione et compositione mathematica,* Rome 1630.

Giles of Rome. Augustinian philosopher and theologian, b. Rome *c.* 1243, d. Avignon; studied at the University of Paris, later taught there. *In libros Posteriororum Aristotelis expositio,* Venice 1488.

Grassi, Orazio. Jesuit mathematician, b. Savona *c.* 1590, d. Rome 1654; professor of mathematics at the Collegio Romano, 1616–1624 and 1626–1628; accepted Galileo's discoveries with the telescope but disputed his views on comets. *Libra astronomica et philosophica,* Perugia 1619; *Ratio ponderum librae et simbellae,* Paris 1626.

Grienberger, Christopher. Jesuit mathematician, b. Tyrol 1561, d. Rome 1636; studied under Clavius at the Collegio Romano, taught the mathematics course there sixteen times between 1595 and 1633; correspondent with Galileo.

Grosseteste, Robert (Lincolniensis). Oxford translator, philosopher, and bishop, b. Suffolk 1175, d. Buckden 1253; studied at the Universities of Oxford and Paris, pioneered in the introduction of Aristotelian thought at Oxford, where he strongly influenced the Franciscans, especially Roger Bacon. *In Aristotelis posteriorum analyticorum libros,* Venice 1514, 1521.

Guevara, Giovanni di. Mathematician and priest, later bishop, b. Naples 1561, d. Teano,

1641; a member of the Clerks Regular Minor and later General of the Order, he served on diplomatic missions for the papacy, was made bishop of Teano in 1627; corresponded with Galileo on the "wheel of Aristotle" and problems relating to the structure of the continuum. *In Aristotelis mechanicas commentarii*, Rome 1627.

Guiducci, Mario. Assistant to Galileo, b. Florence 1585, d. there 1646; studied at the Collegio Romano and at Pisa under Castelli, who introduced him to Galileo; elected consul of the Florentine Academy, 1618; became a member of the Accademia dei Lincei in 1625; published, with Galileo, the *Discorso delle comete*, Florence 1619, which initiated the controversy with Grassi.

Ingoli, Francesco. Lawyer and priest, b. Ravenna 1578, d. Rome 1649; studied law at Padua, where he probably knew Galileo; after ordination served as first secretary of the Propaganda Fidei, founded its famed printing press.

John of Jandun. Averroist commentator on the works of Aristotle, b. Jandun *c.* 1275, d. Todi 1328. *Quaestiones in XII libros metaphysicorum*, Venice 1525, 1553, 1560; *Quaestiones in libros de anima*, Venice 1473, 1561.

Jones, Robertus. English Jesuit, b. 1564, d. 1615; studied and taught at the Collegio Romano, offering the logic course there in 1591–1592. *Organum Aristotelis a Roperto Jones explicatum*, Rome, Biblioteca Casanatense, Cod. 3611 (1592).

Kepler, Johann. German mathematician and astronomer, b. Weil der Stadt 1571, d. Regensburg 1630; studied at Tübingen under Michael Maestlin and one of the first to propagate Copernicus's teachings; worked with Brahe in Prague; had a brief correspondence with Galileo, who mostly ignored his discoveries in planetary astronomy. *Dissertatio cum Nuncio Sidereo*, Prague 1610.

Liceti, Fortunio. Italian philosopher and theologian, b. Rapallo 1577, d. Padua 1657; friend and correspondent of Galileo, studied under Pendasio at the University of Bologna, professor of logic and natural philosophy at the University of Pisa, 1600–1609, then at Padua, 1609–1636, then at Bologna, 1637–1645, and finally professor of medicine at Padua, 1645–1657.

Lorinus, Ioannes. French Jesuit, b. Avignon 1559, d. Dole 1635; professor of philosophy and Scripture at Jesuit colleges in Rome, Paris, and Milan; taught logic at the Collegio Romano in 1585–1586, later a censor librorum for the Jesuit order. *Ioannis Lorini Societatis Iesu Logica*, Vatican City, Biblioteca Apostolica Vaticana, Cod. Urb. Lat. 1471 (1584); *In universam Aristotelis logicam, Commentarii cum annexis disputationibus Romae ab eodem olim praelecti*, Cologne 1620.

Melioratus, Remigius. Italian philosopher, b. Borgo San Sepulchro, d. Pisa 1554; taught philosophy at the University of Padua, 1536–1543, then logic and philosophy at the University of Pisa, 1543–1554. *Expositio in commentum II Averrois libri primi Posteriorum, Quaestio de medio demonstrationis*, Vatican City, Biblioteca Apostolica Vaticana, Cod. Urb. Lat. 1455; *De demonstrationis medio termino*, Lucca 1554.

Mirandulanus, Antonius Bernardi. Italian philosopher and priest, later bishop, b. Mirandola 1502, d. Bologna 1565; studied under Pomponazzi and Buccaferrea at the University of Bologna, where he was later professor of logic and philosophy; bishop of Caserta, 1552–1554. *Institutio in universam logicam, nempe in libros Perihermiensis, Priora, et Posteriora Analytica; in eandem commentarius*, Basel 1549, Rome 1562.

Monte, Guidobaldo del. Friend and correspondent of Galileo, b. Pesaro 1545, d. Urbino 1607; studied mathematics at Padua and then under Commandino at Urbino, 1572–1575.

Mechanicorum liber, Pesaro 1577; *In duos Archimedis aequeponderantium libros paraphrasis*, Pesaro 1588.

Montius, Pamphilus. Italian logician, philosopher, and physician, b. Bologna, d. 1553; studied philosophy at the University of Bologna under Achillini, later taught logic (1510–1515), philosophy (1515–1517), and medicine (1518–1531) there. *Glossemata in libros posteriorum*, Venice 1514, 1521, 1537, 1552.

Neritonensis (Franciscus de Nardo). Dominican philosopher and professor at the University of Padua, fl. 1480, where he probably taught Thomas de Vio Cajetan and Gaspar Contarini; his view on logic, reported by Balduino, were known to Galileo through Jesuit lecture notes.

Nifo, Agostino. Italian philosopher, b. Sessa 1469, d. there 1538; after early studies in Naples, took the doctorate at Padua and taught there until about 1500, then at Naples and Salerno. *Perihermenias interpretatus et expositus*, Venice 1507, 1519, etc.; *In priora analytica commentaria*, Naples 1526, Venice 1543; *In libros posteriorum commentaria*, Naples 1523, Venice 1526, 1548, 1554, 1565.

Nobili, Flaminio. Italian logician, philosopher, and physician, b. Lucca 1533, d. there 1591; studied at the University of Pisa and Ferrara, later taught logic and philosophy at Pisa. *Quaestiones logicae*, Lucca 1562.

Paul of Venice. Augustinian logician and philosopher, b. Udine *c.* 1369, d. Padua 1428; studied at Oxford, taught at Bologna, Perugia, and Siena, where he became rector of the university. *In libros posteriorum magistri Pauli Veneti*, Venice 1491; *Logica parva*, Milan 1473; *Logica magna*, Venice 1481, 1491.

Pererius, Benedictus. Jesuit philosopher and Scripture scholar, b. Valencia 1535, d. Rome 1610; taught logic at the Collegio Romano in 1561–1562 and 1564–1565 and natural philosophy between 1558 and 1566; later taught scripture there. *De communibus omnium rerum naturalium principiis et affectionibus*, Rome 1576.

Philoponus, John. Alexandrian philosopher of the sixth century, a disciple of Ammonius Hermeus; commented on many works of Aristotle, adding Platonic, Stoic, and Christian elements to his own insights. *Commentaria super libros priorum resolutoriorum Aristotelis, commentariae annotationes ex colloqui Ammonii Hermeae*, Venice 1548, 1560; *In libros posteriorum Aristotelis*, Venice 1542, 1560.

Piccolomini, Alessandro. Italian philosopher and priest, later bishop, b. Siena 1508, d. there 1579; studied and taught philosophy at Siena, Padua,and Bologna; archbishop of Patras, coadjutor of Siena. *In mechanicas quaestiones Aristotelis paraphrasis, eiusdem de certitudine mathematicarum disciplinarum commentarium*, Rome 1547, Venice 1565.

Pomponazzi, Pietro. Paduan philosopher, b. Mantua 1462, d. Bologna 1525; studied and taught at the University of Padua, then at the Universities of Ferrara and Bologna. Two anonymous commentaries on the *Posterior Analytics* are attributed to him, one doubtful, Bologna, Biblioteca Universitaria, Cod. 301 (Frati 203), fol. 2r–11v, the other suprious, ibid., fol. 500r–531v (1524).

Porphyry. Neoplatonic philosopher, b. Tyre 234, d. Rome after 301; studied under Plotinus, commented on the works of Plato, composed his *Isagoge*, an introduction to Aristotle's *Organon* that systematized Aristotle's teaching on the predicables and so entered into the medieval and Renaissance Aristotelian university tradition.

Riccardi, Niccolò. Dominican theologian, b. Genoa 1585, d. Rome 1639; joined the Order in Spain; worked for the papal curia in Rome, where he approved for publication

Galileo's *The Assayer* of 1623; made Master of the Sacred Palace in 1629; handled details of censorship for the *Dialogo* of 1632.

Ricci, Ostilio. Mathematician and engineer, b. Fermo 1540, d. Florence 1603; studied mathematics under Tartaglia, taught Galileo privately at Pisa; active in the Academy of Design at Florence; supervised fortifications and hydraulic works; was mathematician to the Grand Duke of Tuscany at his death.

Rocco, Antonio. Philosopher and rhetorician, b. Aquila 1586, d. Venice 1562; studied philosophy at the Collegio Romano, then at Perugia, then at Padua under Cremonini; taught privately at Venice. Published a critique of Galileo's *Dialogo* entitled *Esercitatione filosofiche*, Venice 1633, later annotated by Galileo.

Rugerius, Ludovicus. Jesuit philosopher, b. Florence, taught philosophy at the Collegio Romano 1589–1592, offering the logic course in 1589–1590 and the physics course in 1590–1591. *Commentarium et quaestionum in Aristotelis logicam*, 1589, Bamberg, Staatsbibliothek, Cod. Msc. Class. 62.1–2; *In octo libros Physicorum*, 1590, ibid. 62.2–3; *In quatuor libros De caelo et mundo, In duos libros De generatione et corruptione, In quatuor libros Meteorologicos*, 1591, ibid. 62.4–5.

Scheiner, Christopher. Jesuit mathematician and astronomer, b. Swabia 1573, d. Niesse 1650; studied at Ingolstadt and taught mathematics there, 1610–1616; engaged in prolonged controversy with Galileo over the priority of discovery of sunspots and their nature. *Tres epistolas de maculis solaribus*, Augsburg 1612; *Rosa ursina*, Bracciano 1630.

Scotus, John Duns. Franciscan philosopher and theologian, b. Duns, Scotland, *c.* 1266, d. Cologne 1308; studied at Oxford and Paris and later taught at both universities; the most influential of the Franciscan doctors. Many of the works attributed to him in the *Opera omnia* (Lyons 1639, Paris 1891–1895) are not authentic, including the *Quaestiones utiles super libros priorum analyticorum et posteriorum analyticorum*, Venice 1512, 1520, possibly the works of Marsilius of Inghen.

Simplicius. Neoplatonist commentator on Aristotle, b. Cilicia, *c.* 500, d. after 1533; studied at Alexandria under Ammonius Hermeus, attempted systematically to reconcile Aristotle's teachings with those of Plato. *Commentationes in praedicamenta Aristotelis*, Venice 1550.

Soarez, Cipriano. Jesuit rhetorician, b. Ocana 1524, d. Palencia 1593; taught humanities and Scripture in Jesuit colleges, rector at Braga and Evora; his writings on rhetoric were edited and enhanced by Ludovico Carbone. *De arte rhetorica libri tres ex Aristotele, Cicerone, et Quintiliano deprompti*, Coimbra 1560, 1562, 1575, etc.; Rome 1580; Venice 1588, 1590, etc.

Soto, Domingo de. Dominican philosopher and theologian, b. Segovia 1495, d. Salamanca 1560; studied at the University of Alcalá and the University of Paris, taught at Alcalá and then at Salamanca. *Summulae*, Burgos 1529, Salamanca 1543, 1547, 1554, 1568, 1571, 1582; *In dialecticam Aristotelis, Isagoge Porphyrii, Aristotelis Categoriae, De demonstratione*, Salamanca 1553, 1574, 1583; Venice 1583, 1587.

Suarez, Francisco. Jesuit philosopher and theologian, b. Granada 1548, d. Lisbon 1617; studied at the University of Salamanca, taught theology in Spain and Portugal and at the Collegio Romano from 1580 to 1585. *Compendium logicae universae*, Paris, Bibliothèque nationale, Cod. Lat. 6775 (1585).

Tartaglia, Niccolò. Renaissance mathematician, b. Brescia *c.* 1500, d. Venice 1557; largely self taught, discovered a method of solving cubic equations; pioneered in artillery science;

translated Euclid and Archimedes into Italian, edited the *De ponderositate* of Jordanus de Nemore. *La nova scientia*, Venice 1537; *Quaesiti et inventi diverse*, Venice 1546.

Themistius. Peripatetic philosopher and scholar, b. Paphlagonia 317, d. Constantinople *c.* 388; paraphrased many works of Aristotle to make them available to a wide audience. *Opera omnia*, Venice 1530, 1534; *Paraphrasis in Posteriora Analytica Aristotelis*, Treviso 1481.

Thius, Angelus. Logician and philosopher, fl. *c.* 1547. *Quaesitum et praecognitiones libri Praedicamentorum Porhyriique*, Padua 1547.

Thomas Aquinas, St. Dominican philosopher and theologian, b. Roccasecca *c.* 1225, d. Fossanuova 1274; studied at Naples, Paris, and Cologne, taught at Paris and elsewhere; explained the teachings of Aristotle with such clarity and insight that he became known as "the Expositor." *In Aristotelis librum Perihermenias et Posteriorum Analyticorum expositio*, Venice 1477, 1495, 1496; Paris 1534, etc.

Toletus, Franciscus. Jesuit philosopher and theologian, b. Cordova 1532, d. Rome 1596; studied at Salamanca under Domingo de Soto, taught philosophy at the Collegio Romano, 1559–15; made cardinal, 1593. *Introductio in dialecticam Aristotelis*, Rome 1561, 1565, 1569; Vienna 1562; Venice 1588; *Commentaria una cum quaestionibus in universam Aristotelis logicam*, Rome 1572, Venice 1576, 1581, etc.

Tomitanus, Bernardinus. Logician, philosopher, and physician, b. Padua *c.* 1517, d. there 1576; studied at the University of Padua, taught logic there. *Introductio ad sophisticos elenchos*, Venice 1544; *Animadversiones aliquot in primum librum posteriorum. In novem Averrois quaesita demonstrativa argumenta, Averrois graviores sententiae in primum et secundum libros posteriorum resolutorium*, Venice 1562.

Torricelli, Evangelista. Mathematician and physicist, b. Faenza 1608, d. Florence 1647; studied mathematics with the Jesuits at Faenza and with Castelli in Rome; invented the barometer, was first to explain atmospheric pressure, continued Galileo's work on motion. *De sphaera, De motu gravium, De dimensione parabolae*, all in *Opera geometrica*, Florence 1644.

Trombetta, Antonius. Franciscan philosopher and theologian, b. Padua 1436, d. there 1517; professor of Scotistic metaphysics at the University of Padua, 1476–1511, adversary of Thomas de Vio Cajetan; later bishop of Urbino. *Quaestiones metaphysicales*, Venice 1493, 1502; *Sententia in tractatum formalitatum scoticarum*, Venice 1493, etc.

Ugo Senesis. Italian philosopher and physician, b. Siena, d. 1439; expositor of scientific methodology in the traditions of Aristotle and Galen. *Expositio super libros Tegni Galieni*, Venice 1498.

Vallius, Paulus. Jesuit philosopher, b. Rome 1561, d. 1622; professor of philosophy and theology at the Collegio Romano, teaching logic in 1587–1588; his unpublished notes for this course were plagiarized by Ludovico Carbone and also appropriated by Galileo. *Logica Pauli Vallii Romani ex Societate Iesu, duobus tomis distincta*, Lyons 1622.

Vitelleschi, Mutius. Jesuit philosopher and theologian, b. Rome 1563, d. there 1645; taught philosophy at the Collegio Romano, 1588–1591, logic in 1588–1589; elected General of the Society of Jesus, 1615. Vatican City, Biblioteca Apostolica Vaticana, Cod. Lat. Borgh. 197, *Explicationes in Aristotelis logicam lectae anno 1588 in Collegio Romano*.

Viviani, Vincenzo. Mathematician, Galileo's first biographer, b. Florence 1622, d. there 1703; studied under Piarist Fathers in Florence, resided with Galileo at Arcetri from 1639 to the latter's death, edited the first edition of his works. *De maximis et minimis*, Florence 1659.

Xenocrates. Greek philosopher, b. Chalcedon 395, d. Athens 313; studied under Plato at the Academy, developed his teachings in many writings, none of which has survived, though reports of them are given by Aristotle and Cicero.

Zabarella, Jacopo. Aristotelian philosopher, b. Padua 1533, d. there 1589; studied and taught at the University of Padua, succeeding Tomitanus in the chair of logic, 1564; the foremost commentator on Aristotle in the sixteenth century; *Opera logica*, Venice 1578, Lyons 1587, Basel 1594, Cologne 1597, Treviso 1604, Frankfurt 1608, Venice 1617, etc.; *In duos Aristotelis libros posteriores analyticos commentarii*, Venice 1582.

Zimara, Marc Antonio. Averroist philosopher and physician, b. S. Pietro in Galatina *c.* 1475, d. before 1537; studied at the University of Padua, taught there and at the Universities of Salerno and Naples. *Apostillae*, Pavia 1520–1521; *Contradictiones et solutiones in dictis Aristotelis et Averrois*, Venice 1508, 1516, etc.; *Tabula dilucidationum in dictis Aristotelis et Averrois*, Venice 1537, 1543, etc.

CONCORDANCE OF ENGLISH AND LATIN EDITIONS

English*	Latin**	MS 27***	English*	Latin**	MS 27***
F2.1	1.1	4r	D2.1	46.23	17v
F2.2	3.1	4v	D2.2	49.18	18v
F2.3	4.16	5v	D2.3	54.1	19v
F2.4	6.1	6r	D2.4	58.3	20v
			D2.5	60.22	21r
F3.1	8.5	6v	D2.6	65.3	22r
F3.2	13.13	8r	D2.7	72.1	23v
F3.4	17.1	9r	D2.8	76.1	24r
F3.5	19.20	10r	D2.9	84.1	26v
F3.6	21.16	10v	D2.10	87.5	27r
			D2.11	89.17	27v
F4.1	23.11	11r	D2.12	92.8	28r
F4.2	25.24	11v			
			D3.1	96.22	29r
D1.1	30.17	13r	D3.2	106.19	30v
D1.2	34.9	14r	D3.3	108.24	31r

---

* Paragraph number in the English translation.
** The numeral before the period is the page number in the Latin Edition; the numeral after the period is the line number.
*** The numeral is the folio number; the r and v refer to the *recto* (front) and *verso* (back) sides respectively.

INDEX OF TERMS

INDEX OF NAMES

Boston Studies in the Philosophy of Science

Editor: Robert S. Cohen, *Boston University*

1. M.W. Wartofsky (ed.): *Proceedings of the Boston Colloquium for the Philosophy of Science, 1961/1962.* [Synthese Library 6] 1963
    ISBN 90-277-0021-4
2. R.S. Cohen and M.W. Wartofsky (eds.): *Proceedings of the Boston Colloquium for the Philosophy of Science, 1962/1964.* In Honor of P. Frank. [Synthese Library 10] 1965   ISBN 90-277-9004-0
3. R.S. Cohen and M.W. Wartofsky (eds.): *Proceedings of the Boston Colloquium for the Philosophy of Science, 1964/1966.* In Memory of Norwood Russell Hanson. [Synthese Library 14] 1967   ISBN 90-277-0013-3
4. R.S. Cohen and M.W. Wartofsky (eds.): *Proceedings of the Boston Colloquium for the Philosophy of Science, 1966/1968.* [Synthese Library 18] 1969
    ISBN 90-277-0014-1
5. R.S. Cohen and M.W. Wartofsky (eds.): *Proceedings of the Boston Colloquium for the Philosophy of Science, 1966/1968.* [Synthese Library 19] 1969
    ISBN 90-277-0015-X
6. R.S. Cohen and R.J. Seeger (eds.): *Ernst Mach, Physicist and Philosopher.* [Synthese Library 27] 1970   ISBN 90-277-0016-8
7. M. Čapek: *Bergson and Modern Physics.* A Reinterpretation and Re-evaluation. [Synthese Library 37] 1971   ISBN 90-277-0186-5
8. R.C. Buck and R.S. Cohen (eds.): *PSA 1970.* Proceedings of the 2nd Biennial Meeting of the Philosophy and Science Association (Boston, Fall 1970). In Memory of Rudolf Carnap. [Synthese Library 39] 1971
    ISBN 90-277-0187-3; Pb 90-277-0309-4
9. A.A. Zinov'ev: *Foundations of the Logical Theory of Scientific Knowledge (Complex Logic).* Translated from Russian. Revised and enlarged English Edition, with an Appendix by G.A. Smirnov, E.A. Sidorenko, A.M. Fedina and L.A. Bobrova. [Synthese Library 46] 1973
    ISBN 90-277-0193-8; Pb 90-277-0324-8
10. L. Tondl: *Scientific Procedures.* A Contribution Concerning the Methodological Problems of Scientific Concepts and Scientific Explanation.Translated from Czech by D. Short. [Synthese Library 47] 1973
    ISBN 90-277-0147-4; Pb 90-277-0323-X
11. R.J. Seeger and R.S. Cohen (eds.): *Philosophical Foundations of Science.* Proceedings of Section L, 1969, American Association for the Advancement of Science. [Synthese Library 58] 1974   ISBN 90-277-0390-6; Pb 90-277-0376-0
12. A. Grünbaum: *Philosophical Problems of Space and Times.* 2nd enlarged ed. [Synthese Library 55] 1973   ISBN 90-277-0357-4; Pb 90-277-0358-2

Boston Studies in the Philosophy of Science

13. R.S. Cohen and M.W. Wartofsky (eds.): *Logical and Epistemological Studies in Contemporary Physics*. Proceedings of the Boston Colloquium for the Philosophy of Science, 1969/72, Part I. [Synthese Library 59] 1974
ISBN 90-277-0391-4; Pb 90-277-0377-9

14. R.S. Cohen and M.W. Wartofsky (eds.): *Methodological and Historical Essays in the Natural and Social Sciences*. Proceedings of the Boston Colloquium for the Philosophy of Science, 1969/72, Part II. [Synthese Library 60] 1974
ISBN 90-277-0392-2; Pb 90-277-0378-7

15. R.S. Cohen, J.J. Stachel and M.W. Wartofsky (eds.): *For Dirk Struik*. Scientific, Historical and Political Essays in Honor of Dirk J. Struik. [Synthese Library 61] 1974 ISBN 90-277-0393-0; Pb 90-277-0379-5

16. N. Geschwind: *Selected Papers on Language and the Brains*. [Synthese Library 68] 1974 ISBN 90-277-0262-4; Pb 90-277-0263-2

17. B.G. Kuznetsov: *Reason and Being*. Translated from Russian. Edited by C.R. Fawcett and R.S. Cohen. 1987 ISBN 90-277-2181-5

18. P. Mittelstaedt: *Philosophical Problems of Modern Physics*. Translated from the revised 4th German edition by W. Riemer and edited by R.S. Cohen. [Synthese Library 95] 1976 ISBN 90-277-0285-3; Pb 90-277-0506-2

19. H. Mehlberg: *Time, Causality, and the Quantum Theory*. Studies in the Philosophy of Science. Vol. I: *Essay on the Causal Theory of Time*. Vol. II: *Time in a Quantized Universe*. Translated from French. Edited by R.S. Cohen. 1980 Vol. I: ISBN 90-277-0721-9; Pb 90-277-1074-0
Vol. II: ISBN 90-277-1075-9; Pb 90-277-1076-7

20. K.F. Schaffner and R.S. Cohen (eds.): *PSA 1972*. Proceedings of the 3rd Biennial Meeting of the Philosophy of Science Association (Lansing, Michigan, Fall 1972). [Synthese Library 64] 1974
ISBN 90-277-0408-2; Pb 90-277-0409-0

21. R.S. Cohen and J.J. Stachel (eds.): *Selected Papers of Léon Rosenfeld*. [Synthese Library 100] 1979 ISBN 90-277-0651-4; Pb 90-277-0652-2

22. M. Čapek (ed.): *The Concepts of Space and Time*. Their Structure and Their Development. [Synthese Library 74] 1976
ISBN 90-277-0355-8; Pb 90-277-0375-2

23. M. Grene: *The Understanding of Nature*. Essays in the Philosophy of Biology. [Synthese Library 66] 1974 ISBN 90-277-0462-7; Pb 90-277-0463-5

24. D. Ihde: *Technics and Praxis*. A Philosophy of Technology. [Synthese Library 130] 1979 ISBN 90-277-0953-X; Pb 90-277-0954-8

25. J. Hintikka and U. Remes: *The Method of Analysis*. Its Geometrical Origin and Its General Significance. [Synthese Library 75] 1974
ISBN 90-277-0532-1; Pb 90-277-0543-7

26. J.E. Murdoch and E.D. Sylla (eds.): *The Cultural Context of Medieval Learning*. Proceedings of the First International Colloquium on Philosophy,

Boston Studies in the Philosophy of Science

Science, and Theology in the Middle Ages, 1973. [Synthese Library 76] 1975
ISBN 90-277-0560-7; Pb 90-277-0587-9

27. M. Grene and E. Mendelsohn (eds.): *Topics in the Philosophy of Biology.*
[Synthese Library 84] 1976   ISBN 90-277-0595-X; Pb 90-277-0596-8

28. J. Agassi: *Science in Flux.* [Synthese Library 80] 1975
ISBN 90-277-0584-4; Pb 90-277-0612-3

29. J.J. Wiatr (ed.): *Polish Essays in the Methodology of the Social Sciences.*
[Synthese Library 131] 1979   ISBN 90-277-0723-5; Pb 90-277-0956-4

30. P. Janich: *Protophysics of Time.* Constructive Foundation and History of Time
Measurement. Translated from the 2nd German edition. 1985
ISBN 90-277-0724-3

31. R.S. Cohen and M.W. Wartofsky (eds.): *Language, Logic, and Method.* 1983
ISBN 90-277-0725-1

32. R.S. Cohen, C.A. Hooker, A.C. Michalos and J.W. van Evra (eds.): *PSA 1974.*
Proceedings of the 4th Biennial Meeting of the Philosophy of Science
Association. [Synthese Library 101] 1976
ISBN 90-277-0647-6; Pb 90-277-0648-4

33. G. Holton and W.A. Blanpied (eds.): *Science and Its Public.* The Changing
Relationship. [Synthese Library 96] 1976
ISBN 90-277-0657-3; Pb 90-277-0658-1

34. M.D. Grmek, R.S. Cohen and G. Cimino (eds.): *On Scientific Discovery.* The
1977 Erice Lectures. 1981   ISBN 90-277-1122-4; Pb 90-277-1123-2

35. S. Amsterdamski: *Between Experience and Metaphysics.* Philosophical
Problems of the Evolution of Science. Translated from Polish. [Synthese
Library 77] 1975   ISBN 90-277-0568-2; Pb 90-277-0580-1

36. M. Marković and G. Petrović (eds.): *Praxis.* Yugoslav Essays in the Philosophy
and Methodology of the Social Sciences. [Synthese Library 134] 1979
ISBN 90-277-0727-8; Pb 90-277-0968-8

37. H. von Helmholtz: *Epistemological Writings.* The Paul Hertz / Moritz Schlick
Centenary Edition of 1921. Translated from German by M.F. Lowe. Edited
with an Introduction and Bibliography by R.S. Cohen and Y. Elkana. [Synthese
Library 79] 1977   ISBN 90-277-0290-X; Pb 90-277-0582-8

38. R.M. Martin: *Pragmatics, Truth and Language.* 1979
ISBN 90-277-0992-0; Pb 90-277-0993-9

39. R.S. Cohen, P.K. Feyerabend and M.W. Wartofsky (eds.): *Essays in Memory of
Imre Lakatos.* [Synthese Library 99] 1976
ISBN 90-277-0654-9; Pb 90-277-0655-7

40. B.M Kedrov and V. Sadovsky (eds.): Current Soviet Studies in the Philosophy
of Science. (In prep.)   ISBN 90-277-0729-4

41. M. Raphael: Theorie des geistigen Schaffens aus marxistischer Grundlage. (In
prep.)   ISBN 90-277-0730-8

Boston Studies in the Philosophy of Science

42. H.R. Maturana and F.J. Varela: *Autopoiesis and Cognition*. The Realization of the Living. With a Preface to 'Autopoiesis' by S. Beer. 1980
ISBN 90-277-1015-5; Pb 90-277-1016-3

43. A. Kasher (ed.): *Language in Focus: Foundations, Methods and Systems*. Essays in Memory of Yehoshua Bar-Hillel. [Synthese Library 89] 1976
ISBN 90-277-0644-1; Pb 90-277-0645-X

44. T.D. Thao: *Investigations into the Origin of Language and Consciousness*. 1984
ISBN 90-277-0827-4

45. A. Ishmimoto (ed.): Japanese Studies in the History and Philosophy of Science. (In prep.)
ISBN 90-277-0733-3

46. P.L. Kapitza: *Experiment, Theory, Practice*. Articles and Addresses. Edited by R.S. Cohen. 1980
ISBN 90-277-1061-9; Pb 90-277-1062-7

47. M.L. Dalla Chiara (ed.): *Italian Studies in the Philosophy of Science*. 1981
ISBN 90-277-0735-9; Pb 90-277-1073-2

48. M.W. Wartofsky: *Models*. Representation and the Scientific Understanding. [Synthese Library 129] 1979
ISBN 90-277-0736-7; Pb 90-277-0947-5

49. T.D. Thao: *Phenomenology and Dialectical Materialism*. Edited by R.S. Cohen. 1986
ISBN 90-277-0737-5

50. Y. Fried and J. Agassi: *Paranoia*. A Study in Diagnosis. [Synthese Library 102] 1976
ISBN 90-277-0704-9; Pb 90-277-0705-7

51. K.H. Wolff: *Surrender and Cath*. Experience and Inquiry Today. [Synthese Library 105] 1976
ISBN 90-277-0758-8; Pb 90-277-0765-0

52. K. Kosík: *Dialectics of the Concrete*. A Study on Problems of Man and World. 1976
ISBN 90-277-0761-8; Pb 90-277-0764-2

53. N. Goodman: *The Structue of Appearance*. [Synthese Library] 1977
ISBN 90-277-0773-1; Pb 90-277-0774-X

54. H.A. Simon: *Models of Discovery* and Other Topics in the Methods of Science. [Synthese Library 114] 1977
ISBN 90-277-0812-6; Pb 90-277-0858-4

55. M. Lazerowitz: *The Language of Philosophy*. Freud and Wittgenstein. [Synthese Library 117] 1977
ISBN 90-277-0826-6; Pb 90-277-0862-2

56. T. Nickles (ed.): *Scientific Discovery, Logic, and Rationality*. 1980
ISBN 90-277-1069-4; Pb 90-277-1070-8

57. J. Margolis: *Persons and Mind*. The Prospects of Nonreductive Materialism. [Synthese Library 121] 1978
ISBN 90-277-0854-1; Pb 90-277-0863-0

58. G. Radnitzky and G. Andersson (eds.): *Progress and Rationality in Science*. [Synthese Library 125] 1978
ISBN 90-277-0921-1; Pb 90-277-0922-X

59. G. Radnitzky and G. Andersson (eds.): *The Structure and Development of Science*. [Synthese Library 136] 1979
ISBN 90-277-0994-7; Pb 90-277-0995-5

60. T. Nickles (ed.): *Scientific Discovery*. Case Studies. 1980
ISBN 90-277-1092-9; Pb 90-277-1093-7

Boston Studies in the Philosophy of Science

61. M.A. Finocchiaro: *Galileo and the Art of Reasoning*. Rhetorical Foundation of Logic and Scientific Method. 1980   ISBN 90-277-1094-5; Pb 90-277-1095-3
62. W.A. Wallace: *Prelude to Galileo*. Essays on Medieval and 16th-Century Sources of Galileo's Thought. 1981   ISBN 90-277-1215-8; Pb 90-277-1216-6
63. F. Rapp: *Analytical Philosophy of Technology*. Translated from German. 1981
      ISBN 90-277-1221-2; Pb 90-277-1222-0
64. R.S. Cohen and M.W. Wartofsky (eds.): *Hegel and the Sciences*. 1984
      ISBN 90-277-0726-X
65. J. Agassi: *Science and Society*. Studies in the Sociology of Science. 1981
      ISBN 90-277-1244-1; Pb 90-277-1245-X
66. L. Tondl: *Problems of Semantics*. A Contribution to the Analysis of the Language of Science. Translated from Czech. 1981
      ISBN 90-277-0148-2; Pb 90-277-0316-7
67. J. Agassi and R.S. Cohen (eds.): *Scientific Philosophy Today*. Essays in Honor of Mario Bunge. 1982   ISBN 90-277-1262-X; Pb 90-277-1263-8
68. W. Krajewski (ed.): *Polish Essays in the Philosophy of the Natural Sciences*. Translated from Polish and edited by R.S. Cohen and C.R. Fawcett. 1982
      ISBN 90-277-1286-7; Pb 90-277-1287-5
69. J.H. Fetzer: *Scientific Knowledge*. Causation, Explanation and Corroboration. 1981   ISBN 90-277-1335-9; Pb 90-277-1336-7
70. S. Grossberg: *Studies of Mind and Brain*. Neural Principles of Learning, Perception, Development, Cognition, and Motor Control. 1982
      ISBN 90-277-1359-6; Pb 90-277-1360-X
71. R.S. Cohen and M.W. Wartofsky (eds.): *Epistemology, Methodology, and the Social Sciences*. 1983.   ISBN 90-277-1454-1
72. K. Berka: *Measurement*. Its Concepts, Theories and Problems. Translated from Czech. 1983   ISBN 90-277-1416-9
73. G.L. Pandit: *The Structure and Growth of Scientific Knowledge*. A Study in the Methodology of Epistemic Appraisal. 1983   ISBN 90-277-1434-7
74. A.A. Zinov'ev: *Logical Physics*. Translated from Russian. Edited by R.S. Cohen. 1983   ISBN 90-277-0734-0
      See also Volume 9.
75. G-G. Granger: *Formal Thought and the Sciences of Man*. Translated from French. With and Introduction by A. Rosenberg. 1983   ISBN 90-277-1524-6
76. R.S. Cohen and L. Laudan (eds.): *Physics, Philosophy and Psychoanalysis*. Essays in Honor of Adolf Grünbaum. 1983   ISBN 90-277-1533-5
77. G. Böhme, W. van den Daele, R. Hohlfeld, W. Krohn and W. Schäfer: *Finalization in Science*. The Social Orientation of Scientific Progress. Translated from German. Edited by W. Schäfer. 1983   ISBN 90-277-1549-1
78. D. Shapere: *Reason and the Search for Knowledge*. Investigations in the Philosophy of Science. 1984   ISBN 90-277-1551-3; Pb 90-277-1641-2

Boston Studies in the Philosophy of Science

Boston Studies in the Philosophy of Science

Boston Studies in the Philosophy of Science

Boston Studies in the Philosophy of Science

132. G. Munévar (ed.): *Beyond Reason*. Essays on the Philosophy of Paul Feyerabend. 1991   ISBN 0-7923-1272-4
133. T.E. Uebel (ed.): *Rediscovering the Forgotten Vienne Circle*. Austrian Studies on Otto Neurath and the Vienna Circle. Partly translated from German. 1991
                                                     ISBN 0-7923-1276-7
134. W.R. Woodward and R.S. Cohen (eds.): *World Views and Scientific Discipline Formation*. Science Studies in the [former] German Democratic Republic. Partly translated from German by W.R. Woodward. 1991 ISBN 0-7923-1286-4
135. P. Zambelli: *The Speculum Astronomiae and its Enigma*. Astrology, Theology and Science in Albertus Magnus and his Contemporaries. 1992
                                                     ISBN 0-7923-1380-1
136. P. Petitjean, C. Jami and A.M. Moulin (eds.): *Science and Empires*. Historical Studies about Scientific Development and European Expansion.
                                                     ISBN 0-7923-1518-9
137. W.A. Wallace: *Galileo's Logic of Discovery and Proof*. The Background, Content, and Use of His Appropriated Treatises on Aristotle's *Posterior Analytics*. 1992   ISBN 0-7923-1577-4
138. W.A. Wallace: *Galileo's Logical Treatises*. A Translation, with Notes and Commentary, of His Appropriated Latin Questions on Aristotle's *Posterior Analytics*. 1992   ISBN 0-7923-1578-2
                                      Set (137 + 138) ISBN 0-7923-1579-0

Also of interest:
R.S. Cohen and M.W. Wartofsky (eds.): *A Portrait of Twenty-Five Years Boston Colloquia for the Philosophy of Science, 1960-1985*. 1985   ISBN Pb 90-277-1971-3

Previous volumes are still available.

KLUWER ACADEMIC PUBLISHERS – DORDRECHT / BOSTON / LONDON